Newton in his orchard

Orbital speeds

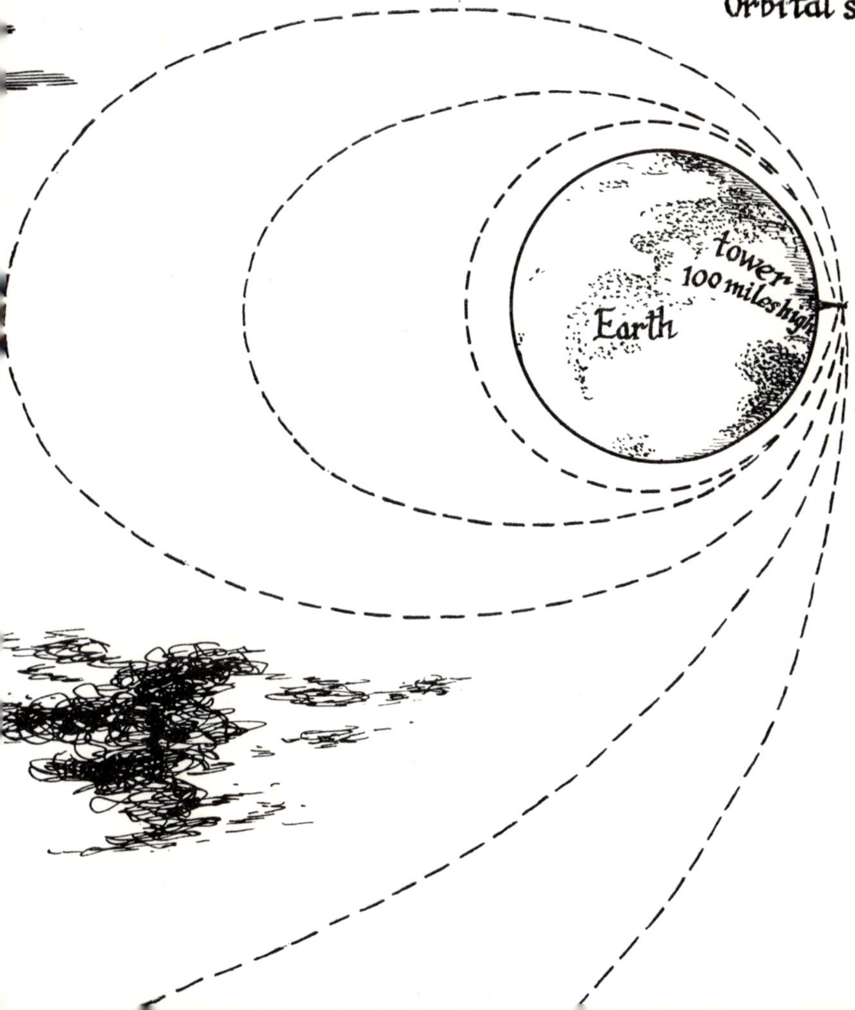

YARDSTICKS OF SCIENCE

River of Time
The Sizes of Things
Taking the Temperature
Feeling the Pressure
Taking Charge

Paul Roberson

A Matter of Mass

ANDRE DEUTSCH

First published 1971 by
André Deutsch Limited
105 Great Russell Street London WC1

Printed in Great Britain by
Ebenezer Baylis and Son Limited
The Trinity Press, Worcester, and London

ISBN 0 233 95997 1

Contents

List of diagrams

List of plates

Preface

Science, once neatly split into the four great divisions of physics, chemistry, biology and mathematics, has become more and more fragmented. Today a bewildering 3 000-odd sub-species are distinguished – from astronomy and aerodynamics to wave-mechanics and zoology.

A great deal of science is pursued for its own sake; with the aim, the romantic would have it, of pushing out the boundaries of knowledge. (Although one eminent British scientist at least prefers to think of scientists 'creeping along the boundaries with a magnifying glass'.) Sometimes, as in the conquest of the Moon or the unravelling of the life processes, many different disciplines are harnessed to a common purpose. But whether practised for its own sake or as part of some grand design, all science advances in tiny steps, the result of painstaking observation and measurement by the scientist.

You are all familiar with some of these measurements – you use them every day, whenever you tell the time, tap the barometer, glance at the speedometer. An American scientist has estimated recently that, by adding to all these casual measurements (perhaps up to fifty a day for every one of us) all those made in industry and by scientists, we arrive at a total of 20 000 million measurements made *every day* in the U.S.A. alone. The figure for Britain, with roughly a third of the population of the U.S.A., might therefore be guessed at 6 000 million.

The *Yardsticks of Science* introduce you to some of the principal properties of matter which the scientist – whether chemist or engineer, materials scientist or plant pathologist – wants to measure. These properties are common to all science, so the authors have drawn their examples without regard to the modern, highly artificial division into 'pure' and 'applied' science. Each book extends across the entire spectrum of its chosen property, from smallest to largest, from coldest to hottest, and so on.

A Matter of Mass and its companion volumes on Time, Size, Electricity, Temperature and Pressure, should help young readers to reach over the barriers that constantly threaten to constrain the scientist himself.

1 Introduction

When you look around you at things, people and animals, and they may be in the room where you are or in the street, in the fields, in the ocean or in the sky, do you ever wonder what they are all made of? Consider all the vehicles, the houses, the enormous variety of goods in stores and factories, intangible things like the wind, the sky, the sun, the moon, the stars, have they all got anything in common? You may know already that living things—animals and plants—have a great deal in common. They all consist of organized arrangements of greater or less complexity, of organic substances which are themselves an expression of the versatility of carbon. Because of its nature, carbon can be the source of an apparently limitless range of chemical substances, some of the most complicated of which can be vitally mixed up in the life process. A number of other substances (elements) can also give rise to complexities but not to the same extent as carbon, at least not on this planet.

This kind of topic has been under consideration by thinking

FIG 1
Burning the candle in the apparatus shown causes an increase in weight as the atmosphere's oxygen is fixed as CO_2 and H_2O, and trapped by the potash and quicklime respectively. The weights and C and H in the gases could be shown to be equal to the loss of weight of the candle.

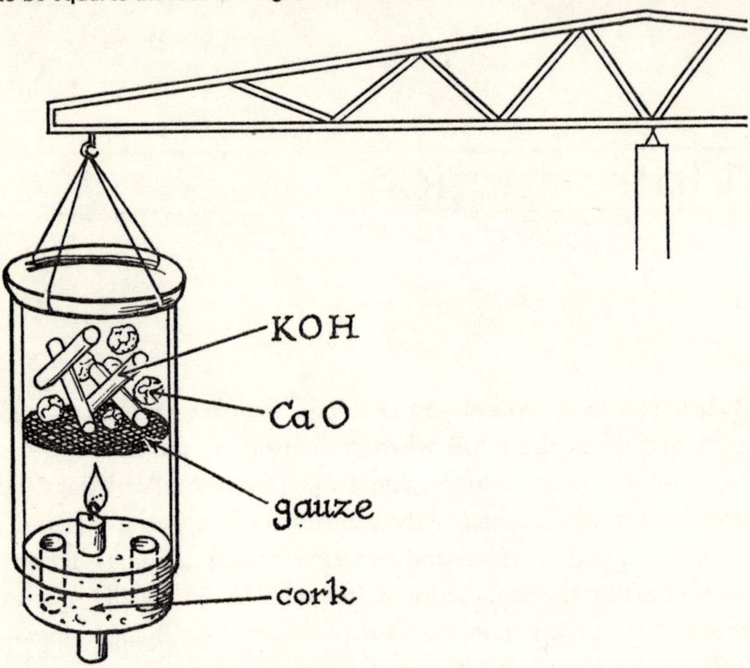

men now for centuries. The early sages decided that the basic stuff from which everything in the universe was made was matter, the primeval stuff that was indestructible and that could not be created. This conclusion, reached by the ancient philosophers, men such as Empedocles (490–430 B.C.) widely quoted by Aristotle, was reached by a process of pure thought. It was completely unverified by experiment, quite unlike the almost identical conclusion reached by Lavoisier some 2 300 years later

FIG 2
The weight of the flask after ignition would be found to be equal to its weight before ignition.

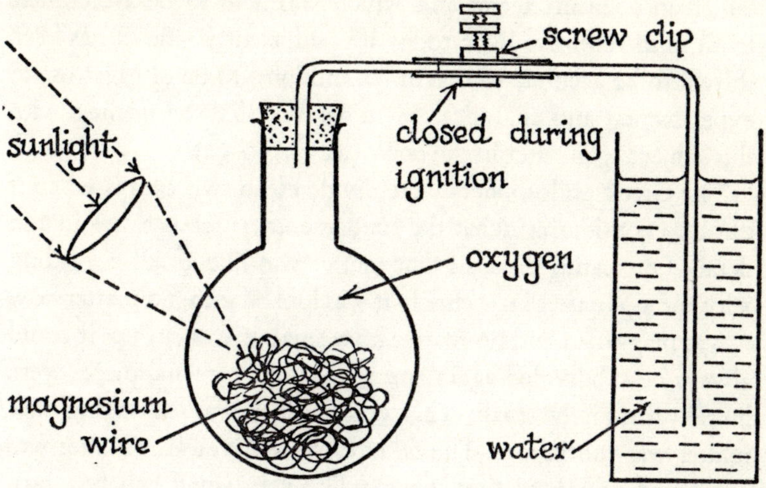

as a direct result of painstaking experiment. The French scientist made a number of careful experiments in which he weighed the ingredients of a number of reactions as accurately as possible before they reacted and carefully collected all the products and weighed them. In all cases he was unable to detect any difference between the sum of the weights of the ingredients and the sum of the weights of the products. Figs. 1 and 2 illustrate this kind of experiment. In the experiment in Fig. 1, the weight does increase, because the oxygen used in combustion is fixed in the CO_2 and water produced by the burning candle. Such experiments were repeated time and time again by many scientists in the years that followed, always with more accurate apparatus, always with more sophisticated experimental techniques and always with the same results—no detectable difference between the sum of the weights of the reactants and the sum of the weights of the products. This process gave rise to the formulation of the Law of the Conservation of Matter which stated the results obtained in scientific terms and which was held to be a complete truth until recently. Even today it is sufficiently true (it has been shown to be accurate to one in ten millions) to be of great use to experimental and analytical chemists. We shall be dealing with this subject again later in the book (see pp. 60–62).

The Greek philosophers were divided into two camps when it came to considering what the fundamental substance was like in detail. One camp thought that matter was like an all-pervading jelly and no matter how closely it was looked into, no matter how it was magnified and no matter how small it was cut up, it could always be subdivided again and again until the final pieces were literally infinitely small. This was another way of saying that matter was continuous. The other camp believed that matter was composed of lots of tiny spheres, like very small ball bearings. Their size was supposed to be very small indeed, too small to be seen even with the most powerful means of magnification. They were imagined to be perfectly elastic and indivisible by physical means which is why they were called atoms (from the Greek meaning 'unable to be cut'). This is of course very close to the view that came to be held by the scientists of the early nineteenth century and indeed to the view held today. Its advantages are that it can offer feasible mechanisms for chemical reactions as well as acceptable explanations for the behaviour of liquids and gases. For example it explains Brownian movement, the behaviour of tiny particles of gamboge suspended in water, illuminated from the side and viewed from above. The particles of the pigment are seen in the field of view of the microscope (Fig. 3) as tiny sparks of light which make random darting movements as if the particles are being knocked about by the particles composing the suspend-

FIG 3
The tiny particles in the illuminated pathway make short darting movements clearly seen in the microscope.

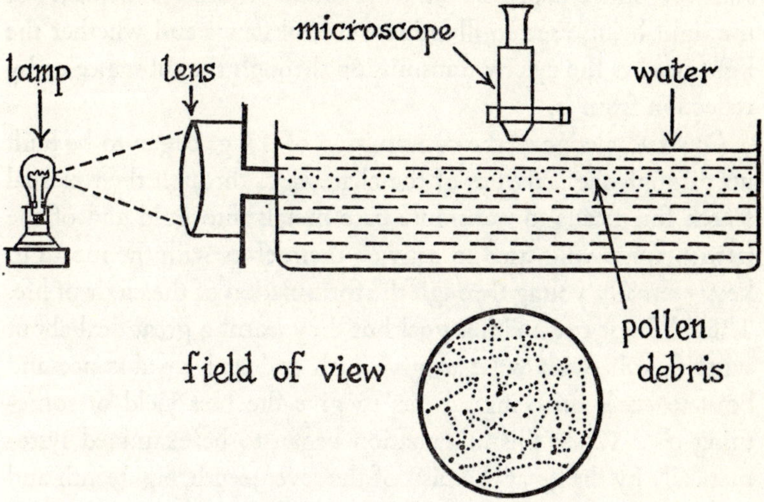

ing fluid, water. The motion was first observed by the botanist Robert Brown in 1827 in suspensions of pollen grains in water. Using a microscope he saw small particles, derived from the pollen, making the same sort of darting motions as were mentioned above. This would happen if the water particles were in ceaseless motion as they would have to be to accord with theories of the composition and behaviour of liquids.

The experience we have of matter is that it comes in many forms. We know it as solid and as liquid by sight and by touch, we know it as a gas by feeling and by smell. We recognize that it can take the characteristic form of crystals and that this means that the components of the crystal are arranged in an orderly and characteristic manner. Other forms of matter have or seem to have no recognizable pattern and they are said to be amorphous (they have no structure). Some forms, when viewed in daylight, appear to be coloured, others appear to be colourless. What is seen very much depends on the eye of the beholder, the quality of the light being used to illuminate the substance and whether the light reaches the eye by transmission through the substance or by reflection from it.

Our knowledge of the composition of things began to be built up through the activities of the alchemists, through their eternal search for means to transmute base metals into gold and of the iatrochemists who tried to provide themselves with the means to keep eternally young through the formulation of the elixir of life. They did not reach either goal but they learnt a great deal about what to mix with what to give such and such a substance and how to cook these ingredients to give the best yield of something else. When this information began to be examined systematically by the great chemists of the seventeenth, eighteenth and early nineteenth centuries, patterns of behaviour emerged which led to the growth and development of the science of chemistry. Starting with Robert Boyle, who began to apply a rigorous scientific method to chemistry, it came to be recognized that there were elementary substances, which could not be broken down by known means into simpler substances, and from which all other substances could be generated. In Boyle's day, only a few elements were known, iron, gold, silver, mercury, copper, zinc, lead and a few others, but gradually, over the years, more and more were discovered until, at the end of the nineteenth century between 80 and 90 were recognized. Incidentally, Boyle must have believed that water was an element as its compound nature was not finally established until nearly 100 years after he died. The process has been made simpler by Mendeleev, the Russian chemist, who in 1869 arranged the chemical elements in order of increasing atomic weight in a periodic table, an arrangement in which elements with similar chemical properties came in the same vertical column. This enabled Mendeleev to predict the total number of elements that there should be between the lightest-hydrogen and the heaviest-uranium. When the table was produced, quite a considerable number of gaps remained to be filled; today the table is complete and has in fact been extended beyond uranium by the new elements that have been created by the nuclear physicists.

From the relatively few building bricks that the 80 or 90 chemical elements represent, all the millions of different chemical substances, known and unknown, in the universe can be constructed. These substances range from 'simple' molecules, and they can be called simple until their detailed structure has to be examined, such as hydrogen or sodium chloride – common salt – to the extremely complicated molecules that are the basis of the genetic code and of life itself. It has been implied above that what may be simple to a chemist is certainly not so to a physicist. This could be due to the chemist being able to treat both atoms and molecules as elastic spheres which somehow stick together to form compounds of greater or less complexity and in doing so release or absorb energy. The physicist on the other hand wants to know why the atoms or molecules stick together, how they manage to do it, how and why they emit or absorb energy, what is it like, why is it in one form in one reaction and in quite another form in a second. As a result of this probing approach, the physicists found out that matter, atoms and molecules, was not immutable, it could be broken down into simpler things and that fundamentally there were three simple building bricks from which all the much more complicated atoms of the periodic table could be constructed. It is true that the construction process is difficult. It probably needs to take place in the most gigantic kind of explosion, bigger by far than the explosion of Krakatoa, much, much bigger even than that which would result from the setting off simultaneously of millions of the largest thermo-nuclear bombs, with which, one must add, the explosion would have a great deal in common. Such a gigantic explosion (known as a supernova) is considered to be the end process in the evolution of a certain type of star (the Sun is not one) in which the atom building processes that have been going on, quietly converting hydrogen into helium, suddenly accelerate when the star collapses. One such explosion was seen to happen about 1 000 years ago (in A.D. 1054) by Chinese astronomers and it has left behind a gigantic and rapidly moving wisp of gas, the Crab nebula. (Plate 1), the prolific source of radio waves that can be detected by radio telescopes. In such explosions, the temperature reaches thousands of millions of degrees centigrade and the pressure many thousands of atmospheres, so that conditions are suitable for the synthesis of all the elements we know on earth from the few simple building bricks we were discussing.

The basic building bricks, and we may assume for the moment they are three in number, are the electron, which is the atom of electricity, the proton which is the positively charged hydrogen atom stripped of its orbital electron and the neutron which is similar to the proton but has no charge. Protons and neutrons are fused together to form the atomic nucleus, different atoms having different numbers of protons in their nuclei. Thus hydrogen has one proton with zero, one or two neutrons in its nuclei, helium has two protons with one or two neutrons and so on up to uranium which has 92 protons and between 142 and 147 neutrons. The physicists discovered that all atoms were similar in the sense that they were all built like miniature solar systems, with a tiny nucleus

FIG 4
The 'solar' atom. A diagram to show the nature of the atom — the nucleus made of protons and neutrons, the planetary of electrons. To give an idea of sizes, if the nucleus were as big as a pea on the centre spot of a rugby football pitch, the electrons would be in orbits beyond the goal posts.

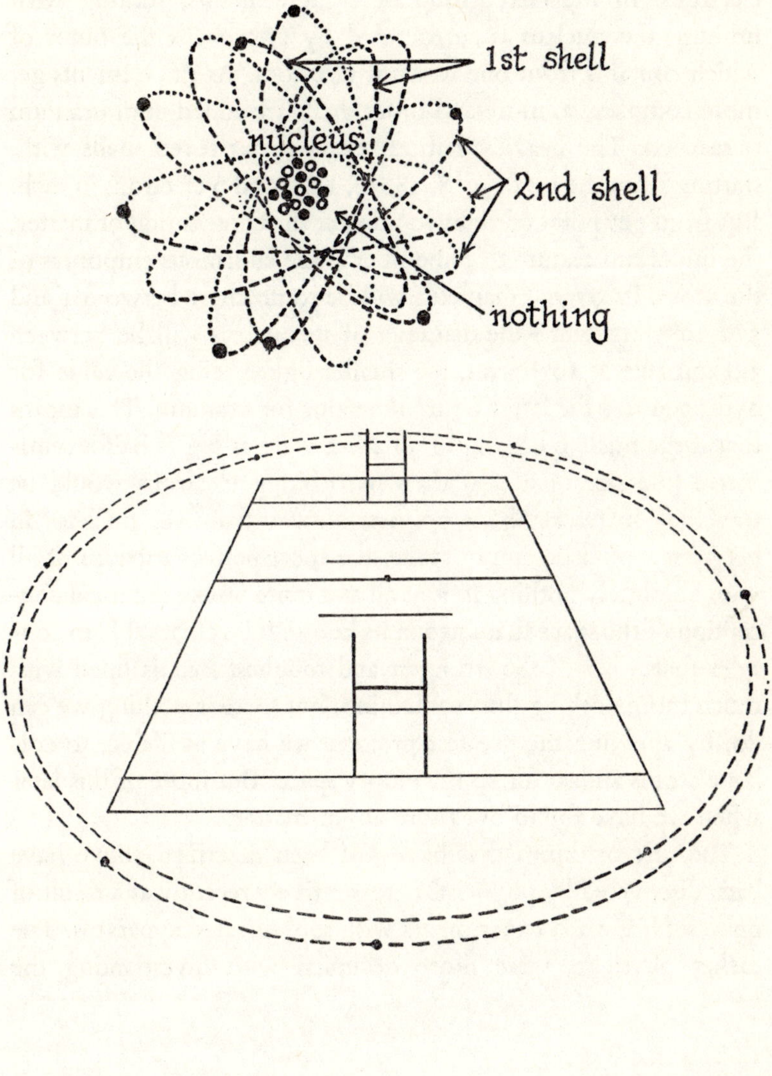

as the Sun, surrounded at a very considerable distance by shells of orbiting electrons. In the simplest atom, hydrogen, the nucleus has one electron in orbit, in helium there are two, also in one orbit, which is then known as a shell as it can contain no more electrons. In the next group of eight elements, starting with lithium, the nucleus is surrounded by two shells the outer of which contains from one to eight electrons. As the elements get more complex, so more and more shells are added until uranium is reached. The heaviest natural element has seven shells with, starting from the inside, 2, 8, 18, 32, 18, 8 and 6 electrons in each. But from our point of view, looking at the constitution of matter, the important feature to remember is the enormous emptiness of the atom. Its overall diameter will be somewhere between 1 and 5×10^{-8} cms while the diameter of its nucleus will be between $3{\cdot}0$ and $18{\cdot}0 \times 10^{-13}$ cms, the smaller figure being the value for hydrogen and the latter figure the value for uranium. This means that if the nucleus (Fig. 4) of an atom were as big as half a centimetre (the size of a pea) then its orbiting electrons would be travelling in tracks about 75 metres away from the nucleus. In between would be empty space, the most perfect vacuum of all with absolutely nothing in it at all and quite unlike the usual conception of the space that astronauts know. It is remarkable to consider that a bar of the strongest and toughest steel is filled with much more nothing than something, but there is nothing we can do, by applying the greatest pressure we have available, to collapse atoms and so fill up the empty space. But more of this later when we have found out more about matter.

The sort of things that have just been described above have been discovered by physicists comparatively recently as a result of quite sophisticated experiments with sophisticated apparatus. The earlier physicists were more occupied with investigating the general properties of matter, how it behaved in given circumstances, what its electrical and magnetic properties were, how strong it was, how elastic, what happened when it was heated or cooled, how much heat it needed to raise it to its melting and boiling point and so on. With an enormous number of different sorts of matter to consider, all the measurements that had to be made kept the scientists busy, particularly as they had to devise techniques for making the measurements with greater and greater accuracy and to find means of using the information when it was obtained. As the information became available it would be used by industry, particularly those industries that were science based, and quite often the commercial performance of a company depended on the accuracy of the information provided by the scientists. One example is an engineering company making electrical generators in which large quantities of copper are used in windings to create magnetic fields. The quality of the performances of the machines will depend on their ability to convert as much as possible of the energy supplied to drive them into electrical energy for distribution. In the nature of things, some of the input energy will be converted into heat by resistance in the windings (and in other ways) and so will not be available for conversion into electrical energy. At the outset, the designer needs to know the resistivity of the copper, so that he can make provision in the design for removing resistive heat. He also needs to know how strong it is and how substances added to improve its strength affect its resistivity and hence heat production. Finally having chosen values for his design, he has to make sure that material having the required properties is actually used in the manufacture so that the machine will really do what the designer says it will. So resistivity and all the other physical properties of copper would have to be measured not only accurately but also by a standard method so that the buyer and the seller could agree that values were true ones.

It should be obvious from all this that one of the fundamental properties of matter that needs to be known by everyone is how much there is of it. The housewife needs to know how much of the kinds of matter called butter, tea, bacon, sugar, beef, etc., there are in the lumps or packets she is being sold. The bullion dealer insists on knowing how much of the kind of matter called gold there is in the ingot he is buying from the prospector. The engineer needs to know how much matter there is in the Apollo capsule, complete with all its equipment and its astronauts sitting on top of the Saturn V rocket waiting to be fired off to the moon. Unless he has this information he won't be able to calculate the trajectory of the craft in space, where it will go to, when to make course corrections or when to fire the braking rockets. Now the property of matter that gives this information, that tells how much matter there is in anything, is the property of matter called MASS.

Mass is the property that tells you that there is more matter in a ten-centimetre cube of lead than in a cube of ice of the same size. Why this should be we will come to in a later chapter but we have already had a clue about this from the way the constituents of matter – atoms – are put together. But we shall also be trying to show you how to measure mass, which is nothing like as easy as it sounds, and of some of the difficulties that crop up when one tries to do this in odd situations such as exist in space.

2 The properties of matter

In the previous chapter, a brief indication was given of the nature of matter, the stuff that makes up everything we see and feel in the world around us. The basic building bricks were the ninety or so elements each of which could, in very special circumstances, be made out of even simpler building bricks – neutrons, protons and electrons. These elements combined in various ways to make up all the great variety of substances that form the rocks, the seas, the living creatures and the artifacts made by man. They all had atomic natures in that they seemed to be made of innumerable and indestructible spheres, all in ceaseless motion of one kind or another. The spheres themselves were always like solar systems of greater or less complexity with a very tiny central sun surrounded by planetary electrons in well-ordered orbits.

The state of the matter would depend to a very great extent on the energy possessed by the component atoms which in turn would depend on their temperature and the kind of binding forces acting between one atom and another. At any one temperature some substances would be gaseous, some liquid, some solid, and, if the temperature were really very high, the matter might be existing in the state of a plasma (Plates 5 and 6). This is a recently discovered state of matter which is probably the most common in the universe. It only exists at very high temperatures and in it all the atoms are separated from their neighbours and all are stripped of all their electronic shells. The plasma thus consists of a mixture of free nuclei and free electrons.

At the other end of the scale, at temperatures only tiny fractions of a degree above absolute zero, the temperature at which no heat energy remains in atoms, all matter can be made to exist in the form of a solid, probably even the most volatile substance known, helium three. At this temperature, in a crystal, where all the 'atoms' occupy precise positions in a spatial arrangement, all the atoms would no longer be vibrating about their nominal positions, they would all be unnaturally still. But as the heat flowed in, the atoms would begin to vibrate and the crystal would take on a more normal appearance. As the substances became hotter, the vibrations would become more and more energetic until the atoms would break away from their fixed positions and the crystal would begin to melt. When melting was complete, all the atoms that were in the crystal would now be free to move about throughout the liquid, some moving with much greater vigour than others. The liquid state, which must be considered with its associated vapour phase, is characterized not only by the freedom of movement of the atoms (or molecules) but also by its surface, which has the properties of an elastic but permeable skin dividing the vapour from the liquid phase. The surface is really a thin zone. It has elastic skin-like properties because the particles within it are only subjected to resultant cohesive forces from below (Fig. 5). In the body of the liquid there are particles having higher energies

FIG 5

Near the surface of a liquid, the cohesive forces within the liquid make the molecules in the surface zone behave as if it were a thin skin. The fast-moving molecules break through the skin to form the vapour about the surface.

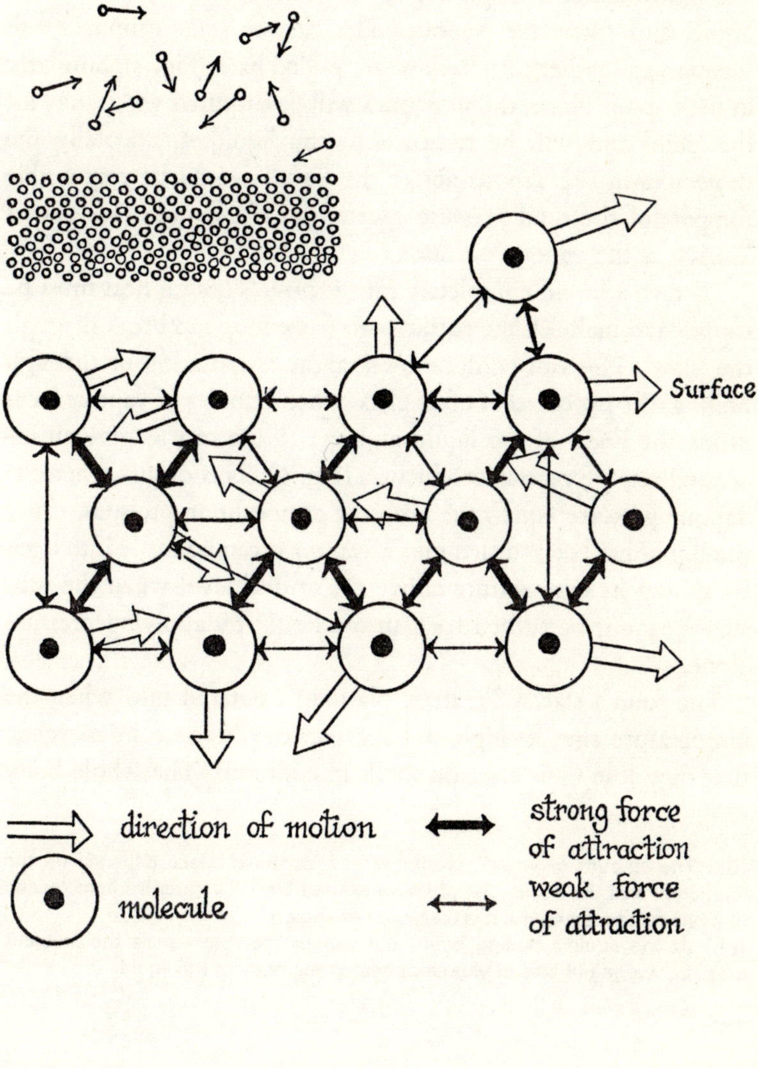

than the average and others having lower energies, these two classes may be considered as groups of particles at temperatures higher and lower than most of their fellows. When the high temperature particles approach the 'skin' zone they are moving fast enough to break through and shoot out into the space above the liquid and form the vapour. The slower ones cannot break through and either turn back or are pulled back (Fig. 5). Similarly in the vapour phase, the slow ones will be captured when they hit the 'skin' and will be returned to the liquid. Incidentally the molecules in the vapour above the liquid would be responsible for part of the total pressure exerted on the liquid. This part is known as the vapour pressure.

To turn a liquid completely into vapour, enough heat must be supplied to make all the particles go fast enough to break through the 'skin'. This can be done by straight evaporation or through boiling, the process that takes place when bubbles of vapour form inside the body of the liquid and rise to the top to give an increased rate of vapour production (Fig. 6). This occurs when the vapour pressure equals the ambient atmospheric pressure. Once the liquid has been turned into a vapour it can be made into a gas by raising its temperature above the critical level when the substance cannot be turned back into a liquid by applying pressure alone.

The fourth state of matter, plasma, is entered into when the temperature rises so high and the particles become so energetic that they lose their electron shells in collisions. The whole body

FIG 6
(6a) The amount of vapour produced in a confined space depends on the temperature of the liquid. The pressure exerted by this vapour is characteristic of liquid and temperature. It is measured as shown.
(6b) At the liquid's boiling point, the vapour pressure equals the ambient pressure. Large bubbles of vapour appear in the body of the liquid.

(6a)

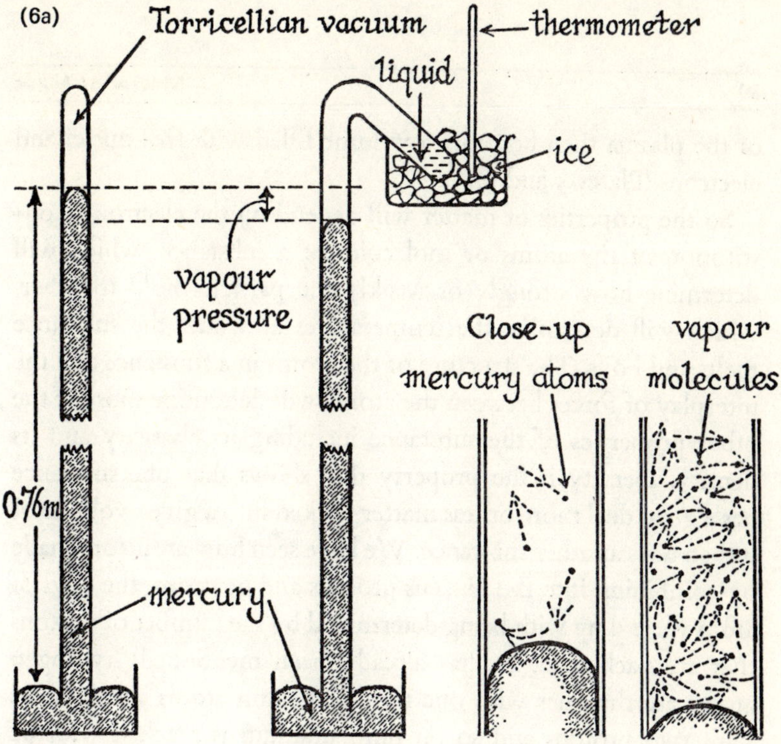
Torricellian vacuum
thermometer
liquid
ice
vapour
pressure
Close-up
mercury atoms
vapour
molecules
0·76m
mercury

(6b)

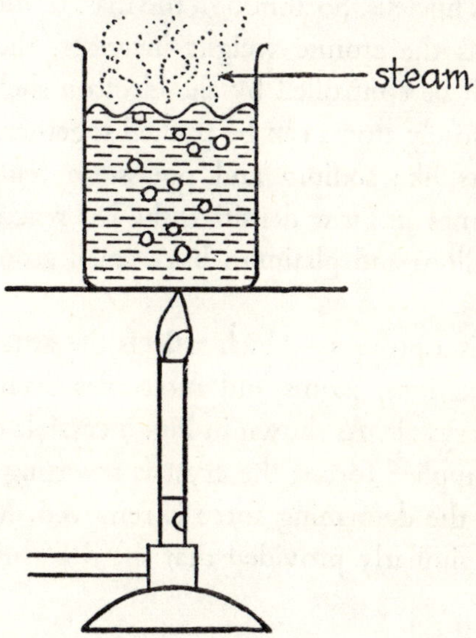
steam

of the plasma then becomes a volume filled with free nuclei and electrons (Plates 5 and 6).

So the properties of matter will depend on the electronic constitution of the atoms or molecules of a substance, which will determine how strongly or weakly the particles hold together, which will determine the temperatures at which the substance melts and boils. The structure of the atoms in a substance and the interplay of forces between the atoms will determine most of the other properties of the substance including its elasticity and its density. Density is the property that shows that one substance has a great deal more or less matter packed into a given volume of it than does another substance. We have seen how atoms are made by combining into the nucleus protons and neutrons, the kind of atom we end up with being determined by the number of protons that are packed in. As has already been mentioned, hydrogen atoms are the ones with one proton, helium atoms are the ones with two protons and so on until uranium is reached with 92 protons in its nucleus. So although the mass of the nuclei of atoms gets larger as the atomic weight increases, the density of the elements will be controlled by the electron shells as they determine how closely atoms can be packed together. The very reactive elements like sodium and potassium tend to have large atomic volumes and low densities, the less reactive elements like osmium, iridium and platinum have small atomic volumes and high densities.

Elasticity is a property which reflects the attraction forces that exist between ions, atoms and molecules even when they are bound into crystals. As shown in Fig. 7 crystals can be deformed by suitably applied forces, the crystals reverting to their original shape when the deforming force is removed. Masses of crystals will behave similarly provided that the deforming force is rela-

FIG 7
The effect of a deforming force, a strain, on a single crystal and on a metal which may be made up of many small crystals.

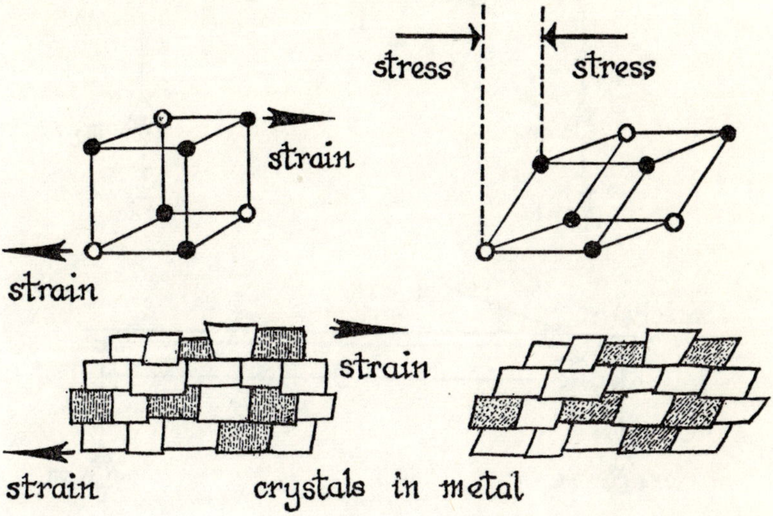

tively small. Large forces will shatter a crystal. Slightly smaller forces cause individual crystals in a mass to slip past one another to give permanent deformations. The resistance shown by a body to deformation is a measure of the strength of the inter-atomic forces.

Turning to a need to determine mass, this must have arisen long before the detailed nature of matter was known. It must have come up about the time thousands of years ago, when trade began to depend on buying and selling instead of plain barter. The method of measuring mass was that of comparing the unknown weight of an article, or animal or ingot, with that of a known weight made up of a series of units or sub-units based on an arbitrarily chosen standard (Fig. 8). The distinction between mass and

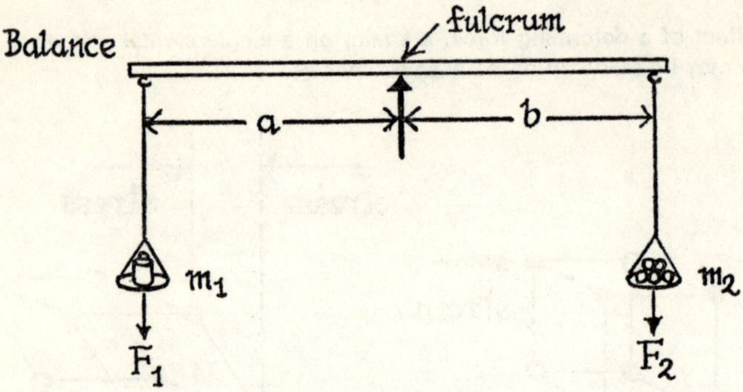

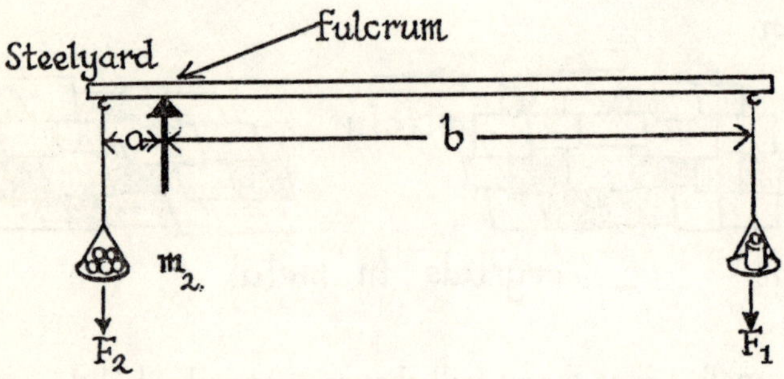

Equal Arm Balance

$$F_1 = \frac{Gm_1 M}{R^2} \qquad F_2 = \frac{Gm_2 M}{R^2}$$

Then at level balance point

$$F_2 b = F_1 a$$

if $b = a$

$$F_2 = F_1$$

$$m_2 = m_1$$

unknown = known

Steelyard

$$F_2 a = F_1 b$$

$$m_2 a = m_1 b$$

$$m_2 = m_1 \frac{b}{a}$$

if a is fixed and m_1 known, mass of m_2 will be measured by distance of m_1 from fulcrum weight came much later. To achieve an accurate and fair result, two things were needed.

1. An accurate balance – one that would set level when equal weights were applied to each side and would move away from the level position when the smallest weight, a small fraction of the mass of the standard, was added to one side.
2. A suitable standard.

Balances having equal arms are known to be of great antiquity, and evidence of their early use and development can be obtained from ancient wall paintings (Fig. 9). The choice of a standard must have been a much more difficult matter and in this connection it is interesting to note the persistence of the word 'grain' through history. It is possible that dried grains of wheat or barley 'taken from the middle of the ear' might have been used to determine the standard basic mass and it is perhaps no coincidence that 180 (three times 60, the common counting base) wheat or barley grains, particularly those from hot countries, come fairly close in weight to a number of standards of mass used by the Indus, Babylonian, Minoan, Egyptian, Sumerian and other ancient civilizations (to the shekel for example). Another possibility is that the grain was the small piece of gold that would be obtained from the primitive mining techniques in use.

The next step was to make a standard of mass from some solid material that was readily available, very durable and readily

FIG 8
The forces of attraction between the Earth and the masses in the balance pans are applied to the ends of the balance beam. If the beam is balanced and the fulcrum at its centre, then the masses in the pans are equal. In the steelyard, the fulcrum is so arranged that there is a simple multiplication factor and the unknown mass is measured by the position at which the known mass is applied on its arm of the lever.

FIG 9
The ancient Egyptian balance had equal arms and a constant radius of curvature pan suspension system.

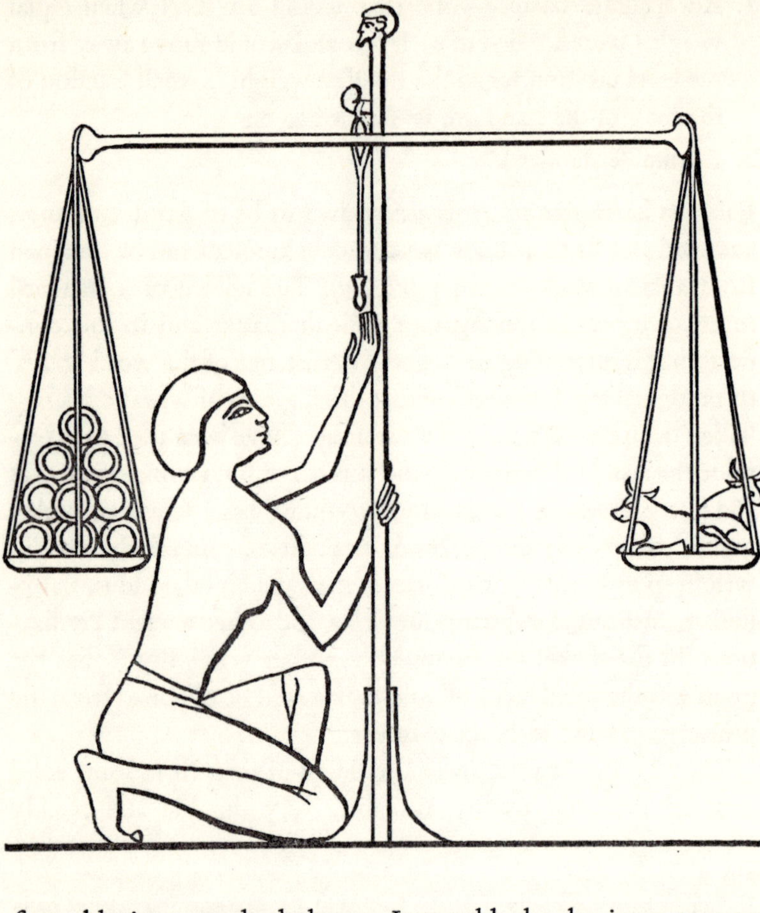

formable into standard shapes. It would also be important to choose a substance that could be so manipulated that its mass could be easily adjusted to match that of the standard. Those concerned with weights and measures in the civilizations mentioned above made up standard series of masses, using a variety of materials including limestone, basalt and metal, and it is remarkable how close their principal standards came to the present-day pound or kilogramme (Fig. 10).

In the United Kingdom, the basic unit of mass, the pound, was derived from that of the Romans as was that of many countries of Europe, and it was subjected to a great deal of variation over the years. The standard was made stable for two hundred years or so by Queen Elizabeth I in 1582 with a series of weights (masses) based on the Troy grain (possibly derived from wheat or barley). In the Troy system, used for bullion today, twenty-four such grains were decreed to be the pennyweight, twenty of which made an ounce of which twelve made the pound Troy. The pennyweight was originally the weight of the silver penny, 240 of which made the pound sterling and weighed one pound Troy. As the coinage gradually became more and more debased, one Troy pound of silver became more and more valuable in relation to the value of the pound sterling. In the merchants system (Haverdepoise or Averdepois or Avoirdupois) again Elizabeth decreed that it should have a pound as the basic standard, that this standard should be equal to 7 000 Troy grains and that it should be subdivided into 16 ounces each of $437\frac{1}{2}$ grains and each ounce into 16 drams of $27\frac{3}{8}$ grains (given approximately but done precisely). Both standards were kept in a safe place in London while a number of copies were made and stored in safe places in fifty-seven towns in England and Wales.

This pound became the sole legal standard in the 1778 Act. The actual standards were cylinders of platinum, a precious and corrosion-proof metal, each measuring about 1·35 inches in height by 1·15 inches in diameter of which six identical copies were made, one stored in London and the other five in various parts of the

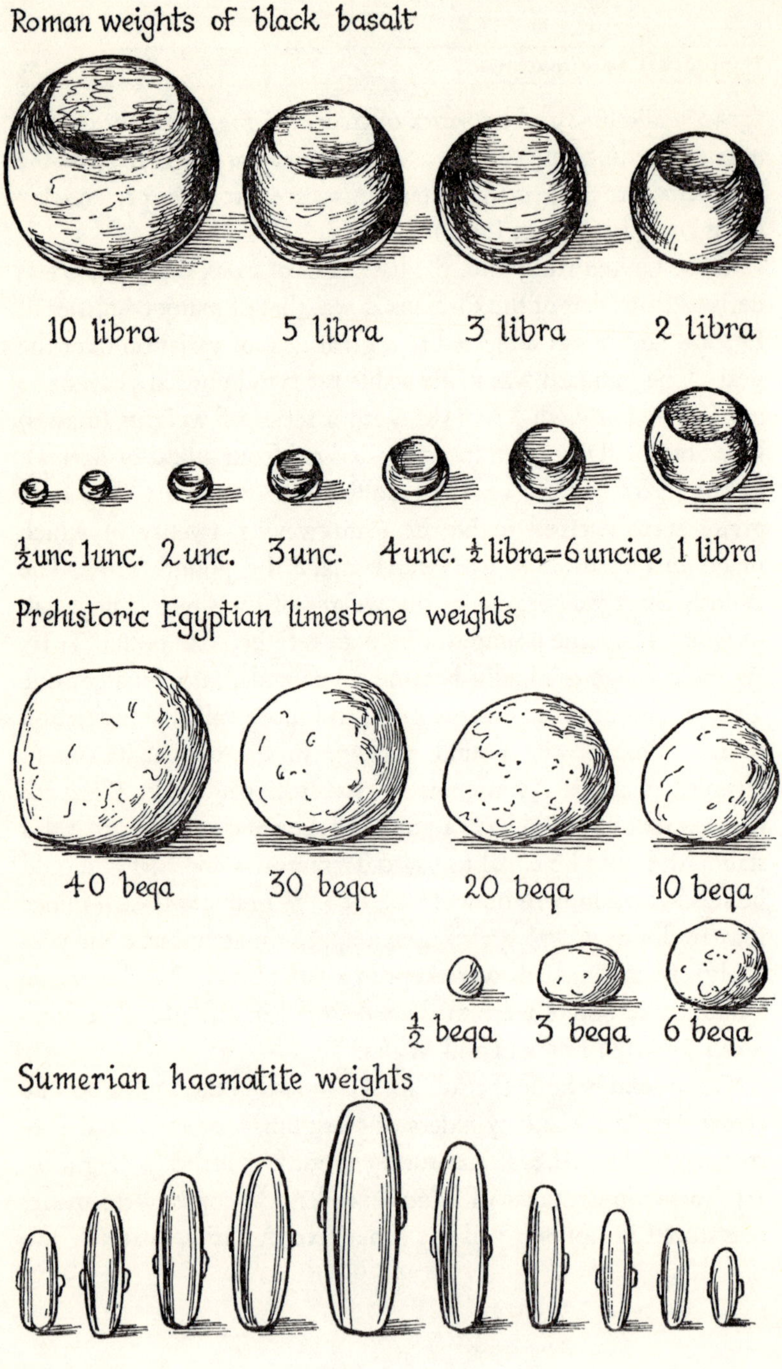
Roman weights of black basalt
10 libra
5 libra
3 libra
2 libra
½ unc.
1 unc.
2 unc.
3 unc.
4 unc.
½ libra=6 unciae
1 libra
Prehistoric Egyptian limestone weights
40 beqa
30 beqa
20 beqa
10 beqa
½ beqa
3 beqa
6 beqa
Sumerian haematite weights

kingdom. These primary Imperial Standard pounds were only brought out on rare occasions for use in comparing the Standards with various substandards. This was done by comparative balancing on a specially constructed high-precision balance originally installed in a basement but now kept at Bushy House, the oldest building of the UK National Physical Laboratory. The balance is housed in a room maintained at extremely constant conditions of temperature and humidity. This balance is capable of comparing two standard pound or kilogramme masses to the order of one millionth of a gramme, that is to the order of one part in one thousand million (Plate 2). In regular comparisons, the standard Imperial pound was found to have decreased in mass quite considerably in the period 1844 to 1922 probably due to excessive wear through too frequent use. The mass then remained constant from 1922 to 1932 and has since increased by a matter of 42 millionths of a gramme possibly due to the accumulation of an invisible film that cannot be removed by brushing with a camel hair brush. In the Weights and Measures Act of 1963, the pound was redefined in metric terms as 0·453592370 kilogrammes.

After the French Revolution, the scientists and philosophers tried to set up rational standards of mass and length (the former based on the latter) based on the then current conception of natural unchanging features. They chose as their standard unit of length, a distance equal to one ten-millionth part of the distance on the Earth's surface between the North pole and the equator (Fig. 11). The kilogramme was then defined as the mass of one cubic decimetre (tenth of a metre = ten centimetres) of pure

FIG 10
Ancient weights made out of many durable but relatively easily shaped and calibrated materials, e.g. basalt, iron, brass, etc.

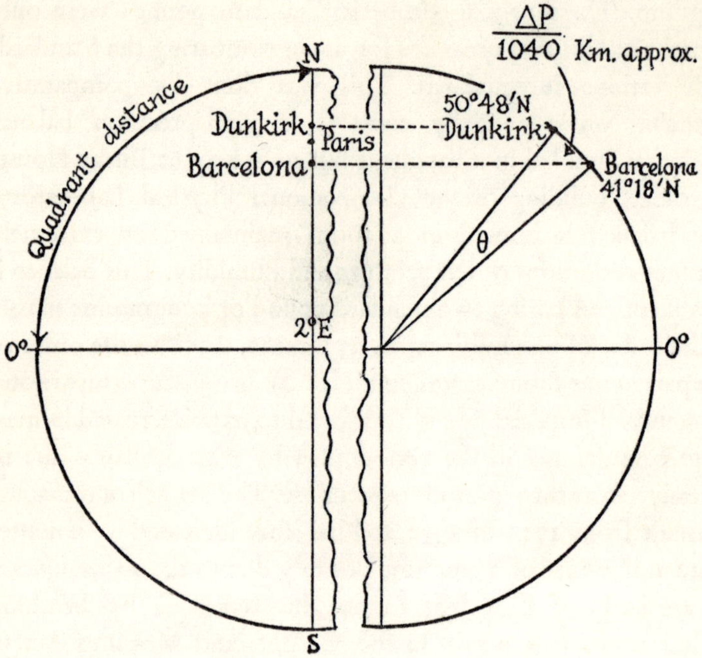
ΔP
1040 Km. approx.
N
Quadrant distance
50°48'N
Dunkirk
Paris
Dunkirk
Barcelona
Barcelona
41°18'N
θ
2°E
0°
0°
S

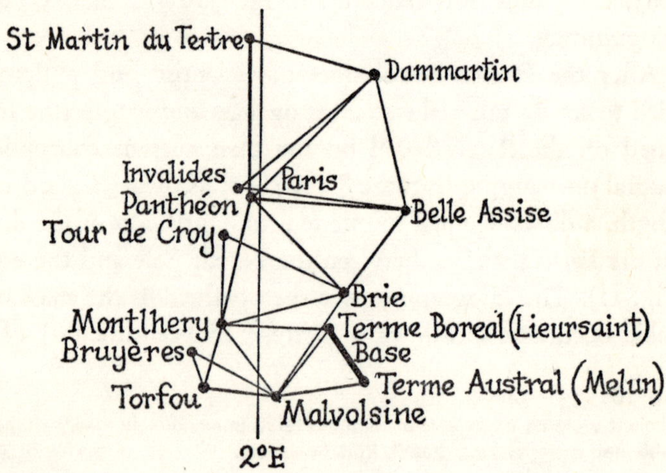
St Martin du Tertre
Dammartin
Invalides
Paris
Panthéon
Belle Assise
Tour de Croy
Brie
Montlhery
Terme Boreal (Lieursaint)
Bruyères
Base
Terme Austral (Melun)
Torfou
Malvolsine
2°E

water at the temperature of maximum density, 4 degrees centigrade. So, theoretically, even if the existing standards of length and mass were destroyed new ones could be created without too much trouble. Practically this is probably not feasible as it was unlikely that the new standards could be guaranteed to be identical with the old within the limits of accuracy required.* Incidentally, the British Standards were destroyed when the Houses of Parliament burned down in 1834. The first permanent presentation of the metric standard mass was the Kilogramme des Archives, a solid cylinder of platinum measuring nearly 4 centimetres high by 4 centimetres in diameter. The original standard, that based on the distance from the North Pole to the Equator, indeed proved to be too inconvenient and too imprecise to be practical. So the specification in natural terms was superseded by a material standard and finally by a 'current standard' of which identical copies verified by the International Bureau of Weights and Measures, Sèvres, are lodged in safe places in most countries of the world. The British copy is No. 18 and its mass has been verified at Sèvres five or six times since it came into existence in 1889 (Plate 3). On all occasions the mass has not deviated more than seven millionths of a gramme from the mean value, better than one part in 100 millions which shows the accuracy and the reliability of the internationally maintained kilogramme standards.

* *Note.* It would now be possible to define a natural mass standard, for example, a definite multiple of the mass of a helium 4 atom.

FIG 11
The metre was originally defined as one ten-millionth part of the distance from the N. Pole to the equator (the Earth's quadrant distance). The quadrant distance was estimated by using triangulation to measure the distance from Barcelona to Dunkirk, which is due north of Barcelona. If the latitudes of these places are known accurately the quadrant distance can be calculated.

FIG 12
The wavelength of light as a length standard.

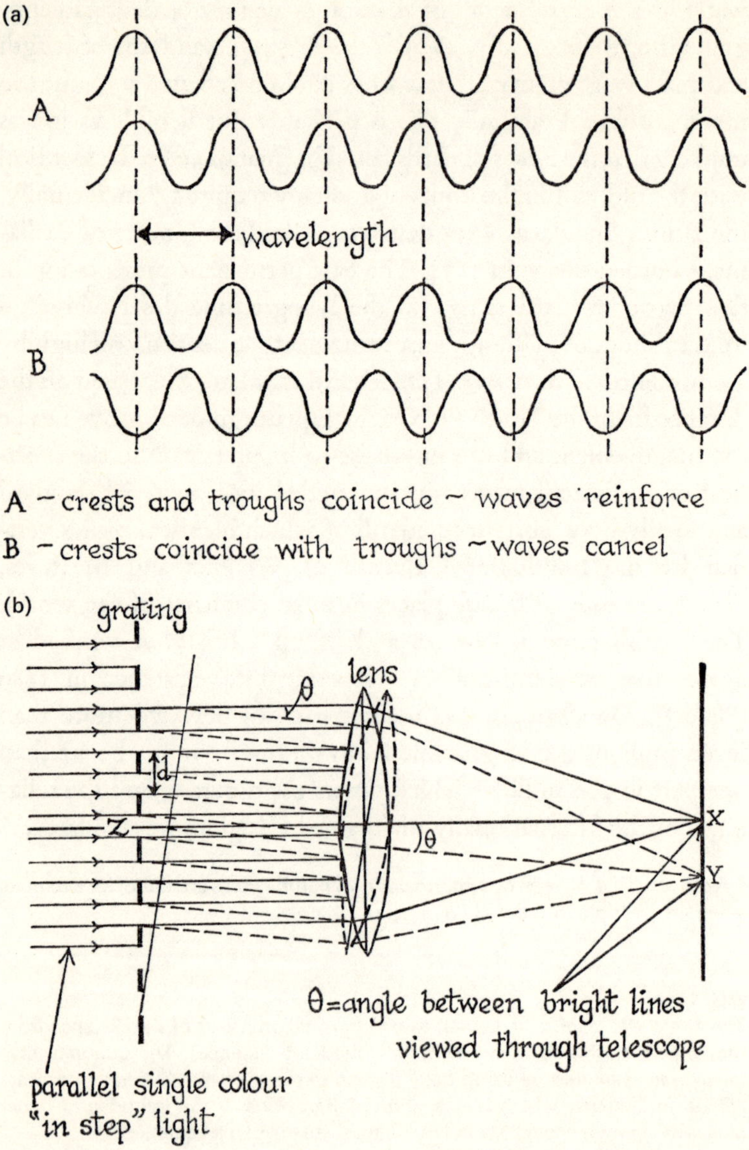

(c)

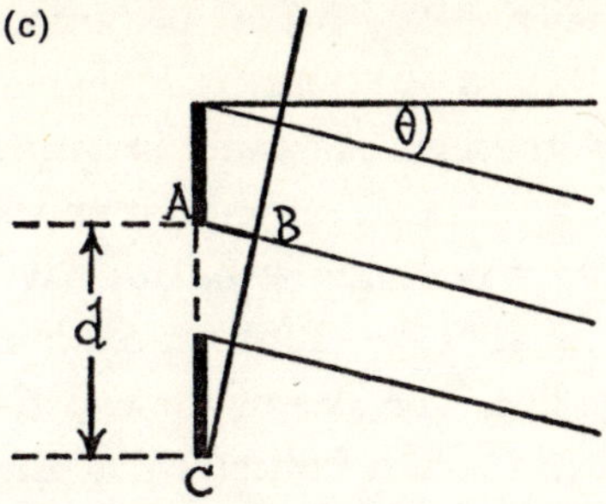

$$AB = 2n\frac{\lambda}{2}$$

$$\frac{AB}{AC} = \sin\theta = \frac{AB}{d} = \frac{XY}{YZ}$$

$$\therefore n\lambda = d \sin\theta_n$$

(a)
The principle of interference. Trains of light waves, all of the same wavelength, interfere when they arrive at a point. If they arrive exactly in step, they reinforce each other (brightness). If they arrive exactly out of step they cancel each other out (darkness).

(b)
The diffraction grating can be used to make waves from a single point source interfere. It consists of a large number of equally spaced, parallel scratches on a glass plate or a metal mirror. It acts like a system of parallel opaque bars across a window.

(c)
The diagram shows how the diffraction grating causes interference fringes to form and how it can be used to measure the wavelengths of light, knowing the width of a bar and the spacing.

Then, for example, one wavelength $= 5{,}000 \times 10^{-8}$ cm
one centimetre $= 20{,}000$ wavelengths

Since the metre and the kilogramme were originally defined and set up and standard metre bars and kilogramme masses constructed and preserved in safe places, it has become customary to measure the lengths of the 'second standard' metre bars by interferometry using monochromatic (single wavelength) light. The red light produced by a cadmium discharge lamp was found to be suitable on account of its purity and homogeneity and the first step was to determine the wavelength of this light in terms of the primary metre standard, the specially shaped bar of platinum-iridium kept in a safe place at the Bureau of Standards at Sèvres. Using accurately made diffraction gratings, calibrated in terms of the standard metre at Sèvres, in experiments carried out under precisely controlled conditions, two wavelength standards have now been accepted internationally, that of the cadmium red line and that of the krypton 86 orange line. The idea of the experiments can be seen from Fig. 12, which also indicates that the measurement of the wavelength depends on being able to measure accurately the distance between the ruled lines on the grating. The value found for the wavelength of the cadmium red line was $6438{\cdot}4708 \times 10^{-8}$ centimetres which means that there are more than 1·553 million wavelengths in a metre. Using another colour light, a very pure orange radiation given off under precisely controlled conditions by the krypton 86 isotope, the metre has now been defined as 'a length equal to 1 650 763·73 wavelengths, in vacuum, of the radiation corresponding to the transition between the levels $2p_{10}$ and $5d_5$ of the atom of krypton 86'.

This now means that having this definition and a supply of krypton 86 anyone anywhere in the world, or in space or on the moon for that matter, can construct a bar which is one metre long to a high order of accuracy. How this could be done is shown in Fig. 13 (a and b). It also means that one could now construct a cubic box whose sides were precisely ten centimetres square at 4 deg. C and at 760 mm of mercury barometric pressure, so that it would contain under these conditions 1 000 cubic centimetres of water which ought to be equal to the mass of one kilogramme. In fact it would be 0·999974 kilogrammes because the kilogramme is now an arbitrary standard constituted by a lump of platinum-iridium kept in the archives. The litre, the accepted standard of volume, is the volume occupied by one kilogramme of

FIG 13

Using a laser to measure lengths accurately. The light rays from a laser are parallel, in step and all of one colour, all one wavelength.

(a) The diagram shows how a laser can be used in an accurate length measuring interference system (known as an interferometer).

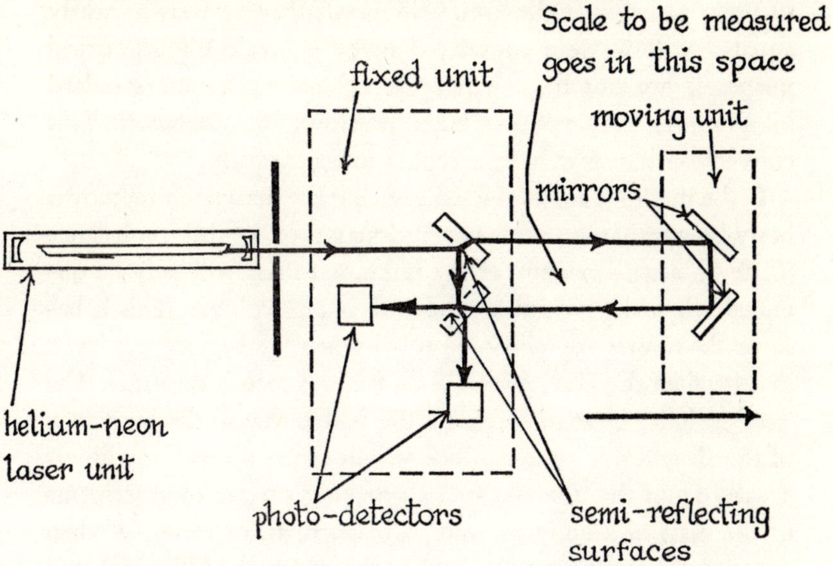

(b) The metre bar used as a standard length in interferometry looks like this. The way it is used to define a metre in terms of a known wavelength of red light is shown in Fig. 13a. The metre bar is placed in the space available for the scale.

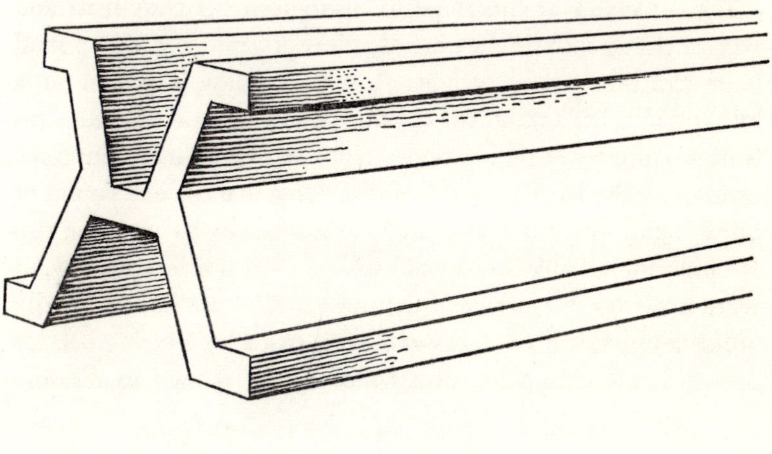

water at 4 deg. C and at an atmospheric pressure of 760 mm of mercury. It is divided into 1 000 millilitres each of which is equal to 1·000028 cubic centimetres. Commonly the two were so nearly equal that they were considered to be equivalent for practical purposes, but not for the purpose of making a new standard kilogramme mass nor for other precision measurements. This convention has now been accepted internationally.

In the methods adopted for measuring the mass of an unknown body by comparison with standard masses on a precision balance (Plate 2), a true measure of the unknown mass will only be obtained when the two masses have the same volume. This is because the masses are said to be equal when the balance sets level, that is when the gravitational pull, etc., on each is the same. The 'etc.' includes upward forces on the bodies due to the buoyancy of the air and this can introduce serious errors if one body differs greatly from the other in size. In making precise measurements of the mass of a body by using a balance, as for example when determining how much a sub-standard mass differs from a standard, allowance has to be made for buoyancy when the materials of which the standards are made differ in density.

It is necessary at this stage to distinguish between mass and weight particularly as they are identical in name, e.g. one pound mass and one pound weight. A unit of mass, a pound or a kilogramme, has a weight of a pound or a kilogramme because this is how commonly understood weight can be defined, the force exerted by the Earth at a certain place on a mass of one pound or one kilogramme. If the measurement is made by balance, the two are indistinguishable and would still be even if the measurement were made on the surface of the moon. They are not necessarily indistinguishable if the measurement is made by a spring balance in which the elongation of a coiled spring is used to measure weight and hence mass. The spring is calibrated by noting how long the spring is when masses of one, two, three, etc., pounds or kilogrammes are attached to its end. It is a property of a spring to extend by double or treble the amount when *forces* of one, two or three times the original force are applied to the end. So the 'balance' will measure the mass of an unknown object only at the place where it was calibrated in terms of standard masses, or at places where the attraction due to gravity is the same as at the place of calibration. If the spring balance were calibrated at sea level on the earth, the balance would not be able to give an accurate measure of mass if it were used on the Moon or on top of a high mountain, or at a place where the pull of gravity were different. As an interesting example of the effects of variations in gravity, during the recent American expeditions to the Moon, great concern was caused at the Texas Control Centre when the first manned spacecraft was late in reappearing from behind the Moon. This anomaly, with some others, was eventually traced to unexpectedly high local values of gravity caused by mascons (short for mass concentrations) below the lunar surface. Such mass concentrations are thought to be created by meteorites crashing into the Moon – meteorites either of high density or so energetic as to compact the Moon substance.

3 Measuring mass

It was stated earlier that mass is the property of matter that tells us how much matter there is in the piece or lump of matter under consideration. So, to measure the matter in a lump, we have to measure how much mass there is in it and this measurement we cannot make directly, that is in the same way as measuring a piece of string. All that can be done is to observe how the piece of matter would behave in certain circumstances and then deduce from the behaviour how much mass there is, at least with reference to the place and circumstances in which the measurement was made. The reason for this peculiar statement will become clear when we know more about the relationship between matter and energy.

Among the many properties of matter, two that can be used to measure most precisely how much mass is present are: the property matter has of attracting to itself other matter – the property of gravitational attraction and the property of inertia – the property that makes matter continue in a state of rest or of uniform motion in a straight line unless acted upon by an external force.

These two properties can be said to relate to two different kinds of mass, gravitational and inertial, and they have to be shown to be identical.

Although a knowledge of falling must go back to the beginnings of man's life on earth, and the knowledge is so old that it is almost instinctive, knowledge about what caused falling is relatively recent and originated with man the thinker – whenever he first appeared. Many philosophers concerned themselves with the problem and in the seventeenth century, when Galileo was said to have performed his experiments with bodies dropped from the leaning tower of Pisa, the concept of earthly gravitation had long been commonly accepted. But by tradition, the full significance of falling was not realized until it struck Sir Isaac Newton when observing a falling apple in the garden of his mother's house in Norfolk. In this flash of inspiration, Sir Isaac realized that the same force that pulled the apple to the ground, the force of attraction between the Earth and the bodies on it or near it, was the force that caused the Moon and the planets to circle round their parent body rather than go off undeviated into space. When a stone is tied to a string and circled round your head, the string supplies the force needed to keep the stone in a circular path. If the string breaks, the stone flies off in a straight line, at a tangent to its previous path. The force supplied by the string is similar to gravitational force.

The four great astronomers, Copernicus and Tycho Brahe in the sixteenth century and Kepler and Galileo in the seventeenth century, had between them solved all the current major problems of the Solar System save one, what kept the planets fixed in their orbits as they revolved around the Sun (or the Earth if you

FIG 14
The planet's orbit is exaggerated to show that it is an ellipse with the Sun at one focus. It also shows that the planet must move fastest at perihelion if the second of Kepler's Laws is to be obeyed.

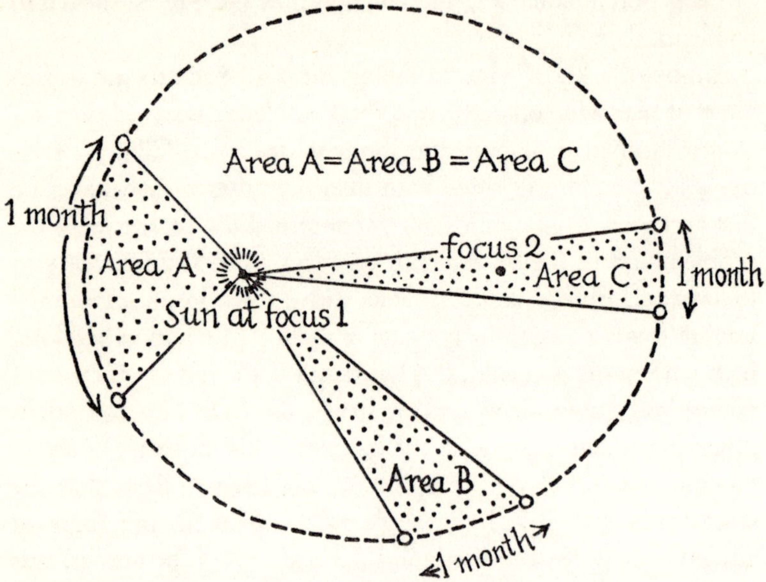

preferred it). Copernicus, as a result of his observations, was responsible for the notion, against currently accepted religious thought, that the Sun and not the Earth was the centre of the Solar system and that the planets Mercury, Venus, Earth, Mars, Jupiter and Saturn all revolved about the Sun in circular orbits held in position by invisible crystal spheres. Tycho Brahe, still believed that the Earth was the centre of the Universe and postulated that although the planets revolved about the Sun, the Sun, with the planets, revolved about the Earth. His major contribution was however the painstakingly accurate observations that he made of the movements of the Heavenly bodies, data that was to prove invaluable to the pupil of his declining years, Johannes Kepler. Kepler, a German like his master, had the job of completing Brahe's tables of the motions of the planets and, as a result of studying the whole system as then known, was able to support the Copernican theory by showing how the Solar system could make sense if all the planets, including the Earth, revolved around the Sun. From Brahe's observations, which he completed, Kepler was able to formulate the three following laws (Fig. 14):

1. The planets move about the Sun in elliptical orbits with the Sun at one focus of the ellipse.
2. The imaginary line (the radius vector) joining the planet to the Sun sweeps through equal areas in equal times.
3. The square of the time of revolution of any planet about the Sun is proportional to the cube of its mean distance from the Sun.

These laws, although they described planetary movement accurately, could give no hint of a mechanism for holding the planets in their orbits, and Kepler had still to resort to the idea of the crystal sphere. Newton, with his theory of universal gravitation, was able to dispense with the crystal sphere and provide a perfectly satisfactory explanation of the way the planets, and for that matter a great number of other things, moved in the heavens and elsewhere.

The fourth of the great astronomers, Galileo Galilei, better known as Galileo, used the newly invented telescope to make a series of discoveries about the planets that settled, at least for him, the problem of the solar system once and for all in favour of Copernicus. In his early years he discovered the time-keeping properties of the pendulum by watching, it is said, the swing of the lamps in church, and by establishing the constancy of the period by timing swings with the aid of his pulse. He also

FIG 15

The Acceleration-due-to-Gravity, g_1, at a point in space can be determined from the inverse square law, i.e.:

$$g_1 = g_0 \cdot \frac{d_0^2}{d_1^2}$$

where g_0 = acceleration at Earth's surface
d_0 = radius of Earth
d_1 = distance Earth centre to point in space

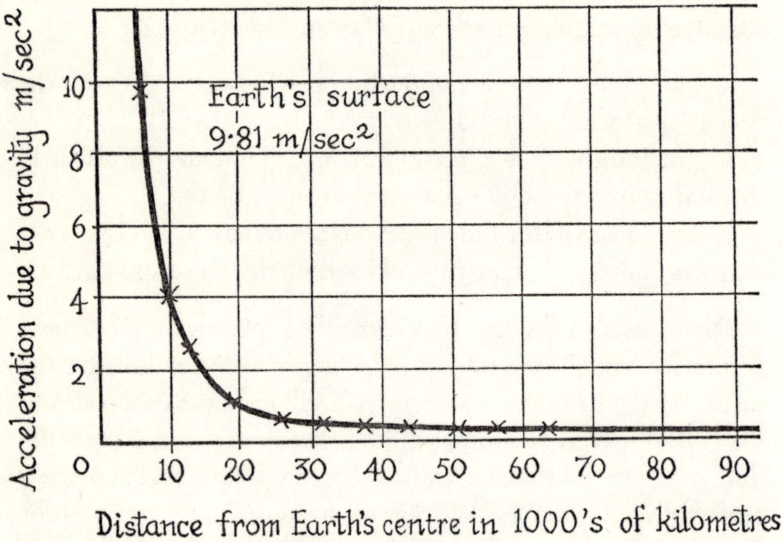

observed that the period did not depend on the amplitude of the swing. This was a valuable discovery in an age when accurate means of measuring time were badly needed by the great navigators. His observations of the heavens with a home-made telescope led to discoveries which included the four satellites of Jupiter in 1610, mountains on the Moon, the rings of Saturn, the phases of Venus and the movement of sunspots which showed that the Sun also rotated about its axis. Galileo also resolved the Milky Way into a huge number of individual stars.

Sir Isaac Newton, having, it is said, realized the significance of a falling apple, applied his law of Universal Gravitation to Kepler's pronouncements and was able to explain planetary movement. His own statement of the law of attraction runs: 'Every particle of matter in the universe attracts every other particle with a force whose direction is that of the line joining the two, and whose magnitude is directly proportional to the product of the masses divided by the square of their distance from each other.' (The corollary to this – acceleration in falling only dependent on distance *not* on mass – is shown in Fig. 15.) The need for the crystal spheres to keep the planets in their courses was eliminated by the concept that the force of attraction could now act like the piece of string holding the circling stone into its path, a force that overcame the tendency of the planet to fly off along the tangent. Gravitation could also explain Kepler's other laws, including the elliptical orbits, and it also provided a means of measuring the masses of the Moon, the planets and the Sun. Incidentally it showed that the techniques that had been in use for thousands of years for measuring out quantities of gold were sound and still good enough to measure out the quantity of butter we wish to purchase in the grocery store today.

In mathematical terms Newton's Law of Universal Gravitation can be expressed as:

$$F = \frac{G M m}{d^2}$$

where F = the force of attraction between two masses m and M, distance d apart and G = the constant of gravitation.

Apart from its use in confirming the soundness of the methods evolved on earth for measuring mass by comparison, methods that depend on balance and moments (see Fig. 9, p. 34), the formula is of little use for determining the mass of the Earth, the

Moon, the Sun and all the other planets unless values can be given to the masses and the distances and in particular to the constant G. The first experiment to determine the value of G was planned by the Rev. John Mitchell in the mid-eighteenth century and carried out on his death by Henry Cavendish at Cambridge using the apparatus sketched in Fig. 16. Simply the apparatus consisted of a rigid bar attached firmly at its middle point to a long metal filament supported from a torsion head, a device that would allow the suspension to be twisted by a precisely measurable amount one way or the other. At each end of the rigid bar was a lead ball, of precisely the same mass and dimensions as its companion, about one inch in diameter. By means of a ray of light reflected from a small concave mirror attached to the suspension filament close to the rigid bar, both the rest position and the deflected positions of the bar could be measured. With the whole apparatus encased in a draught- and heat-proof enclosure, the small balls were deflected first one way and then the other by bringing up to the small balls a pair of large diameter lead balls, one above (as we look at the diagram) the left-hand small ball, one below the right-hand small ball. In this case, the swinging system would be deflected slightly in a clockwise direction. By changing the large balls over, the opposite deflection could be brought about and the torsional force needed to overcome it measured by the torsion head.

FIG 16
The Cavendish experiment to determine G. The small lead balls are mounted on a light beam suspended at its centre on a very thin wire (or quartz filament) from a torsion head. The large lead balls are mounted on a heavy beam that can be rotated in a horizontal plane about its centre. In the experiment the force of attraction between small and large balls is measured by the torsion force that has to be applied to correct it. This is done with the large balls first in one side, then the other.

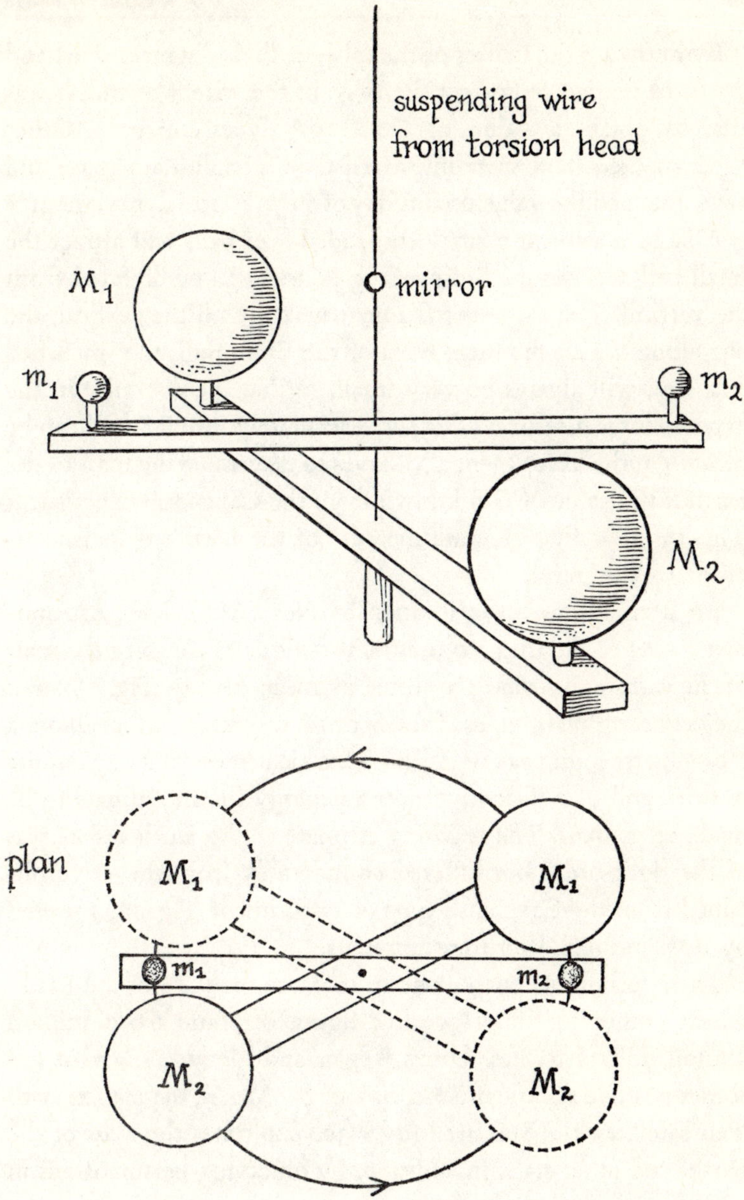
suspending wire
from torsion head
mirror
M_1
m_1
m_2
M_2
plan
M_1
M_1
m_1
m_2
M_2
M_2

By knowing the masses of the spheres, their distances apart and the force needed to correct the twist in the wire, Cavendish was able to obtain a value of $6{\cdot}6 \times 10^{-8}$ dynes cm^2 gr^{-2}. Other experimenters have since improved the Cavendish apparatus and have obtained the value used today of $6{\cdot}670 \times 10^{-8}$ dynes cm^2 gr^{-2}

A large mass near a small suspended lead ball, will attract the small ball and cause its suspending filament to be diverted from the vertical. The angle the filament makes with the vertical, the line along which the huge mass of the Earth pulls the small ball (Fig. 17), will always be very small, so that the difficulty in the experiment is measuring this angle accurately. Providing it can be measured, the experiment also serves to determine the mass of the Earth if the value of G is known from the Cavendish experiment (Fig. 16). The first accurate measure of the Earth's mass is attributed to Cavendish.

An attempt was made in 1772 by Nevil Maskelyne, Astronomer Royal of the time, to use the technique to measure the mass of the earth and hence determine its mean density (Fig. 18). For the deflecting mass, he used the Scottish mountain Schiehallion, a 1 048-metre high peak in Perthshire which was sufficiently regular in form and of sufficiently known geology for an estimate to be made of its mass. The resulting estimate of the Earth's mass was of the right order but differed considerably from the value obtained from the Cavendish type of experiment (Fig. 16, 17) used by Boys in much later measurements.

These techniques enable one to measure the mass of the Earth, which comes to the staggering figure of some 6 000 million million million tonnes. Using Kepler and Newton's laws it becomes possible to estimate the mass of the Moon, the planets with their satellites, the Sun itself (over 300 000 times the mass of the Earth) and other stars. In addition, by observing perturbations in

FIG 17
Measuring the mass of the Earth directly. If a small lead ball (a few grammes) is suspended by a thin cord near a large lead ball (one tonne) so that the line joining their centres is horizontal, the small ball will be attracted to the large one and the suspending cord pulled out of the vertical. If the angle between the cord and the vertical can be measured, the mass of the Earth can be calculated.

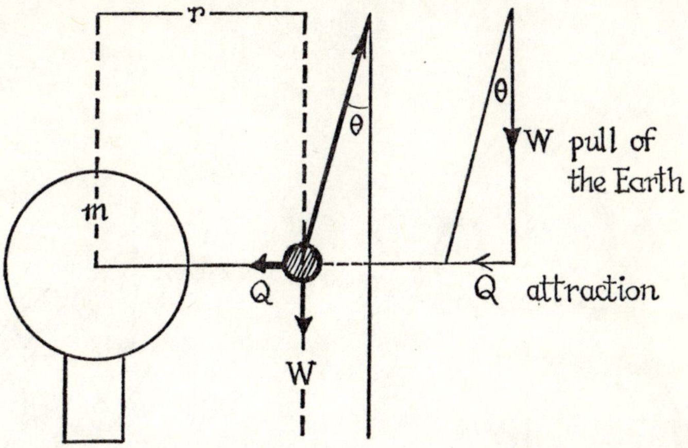

m = spherical mass of lead = 1 ton = 1,016,360 grammes

$$\text{Tan}\ \theta = \frac{mR^2}{Mr^2} \quad \therefore \text{ mass of Earth} = M = \frac{mR^2}{\tan\theta . r^2}$$

$$\text{If } \tan\theta = 8 \times 10^{-9}, \quad M = 6{\cdot}0 \times 10^{21} \text{ tonnes}$$

the motions of planets or stars, it has been possible to discover by calculation that new and unseen planets should be about, and to know before they had been found what they were like, how massive, how far from the Sun and how long were their periods of rotation.

The second method for estimating mass depends on the property it has of continuing in its state of rest or of uniform motion in a straight line unless acted upon by a force that could divert the matter from the course upon which it was set. So if one had a stationary piece of matter and one brought to bear on it a

FIG 18
Astronomer Royal, Nevil Maskelyne, used the Scottish Mountain, Schiehallion, in an experiment similar to the one mentioned in Fig. 17. He estimated the mass of the mountain by knowing its geology and measured the deflection of the cord astronomically.

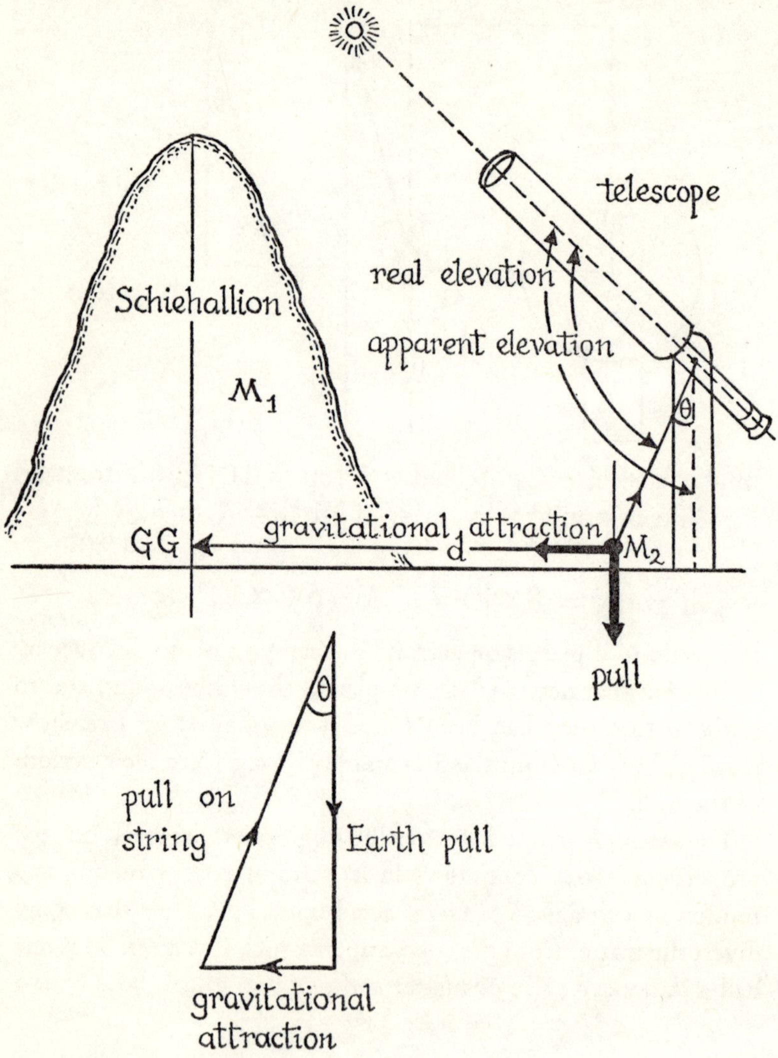

Plate 1. In AD 1054 Chinese astronomers observed a colossal stellar explosion in the heavens, the kind of explosion known today as a super nova. The gaseous residues of this explosion are known as the Crab nebula. They are streaming away from the Earth at 1 000 km. per sec. and are a prolific source of radio waves.

Plate 2. This balance, installed in a controlled environment room at the National Physical Laboratory, is used to compare standard and sub-standard masses, such as kilogrammes, to an accuracy of 1 part in 10^9, i.e. to 1 microgramme in 1 kilogramme.

Plate 3. Standard Kilogrammes are cylinders of about 4 cms diameter and 4 cms height made of a hard corrosion-proof alloy — platinum iridium. No. 18, the standard shown, is one of the British main standards. The universal primary standards are kept in France.

Plate 4. A Beta-ray thickness gauge. The loss of intensity in a beam of Beta-rays from a radio-active source can be used to control the thickness of steel sheet (see Fig. 19). Such a gauge, in use in a steel sheet works, is shown here.

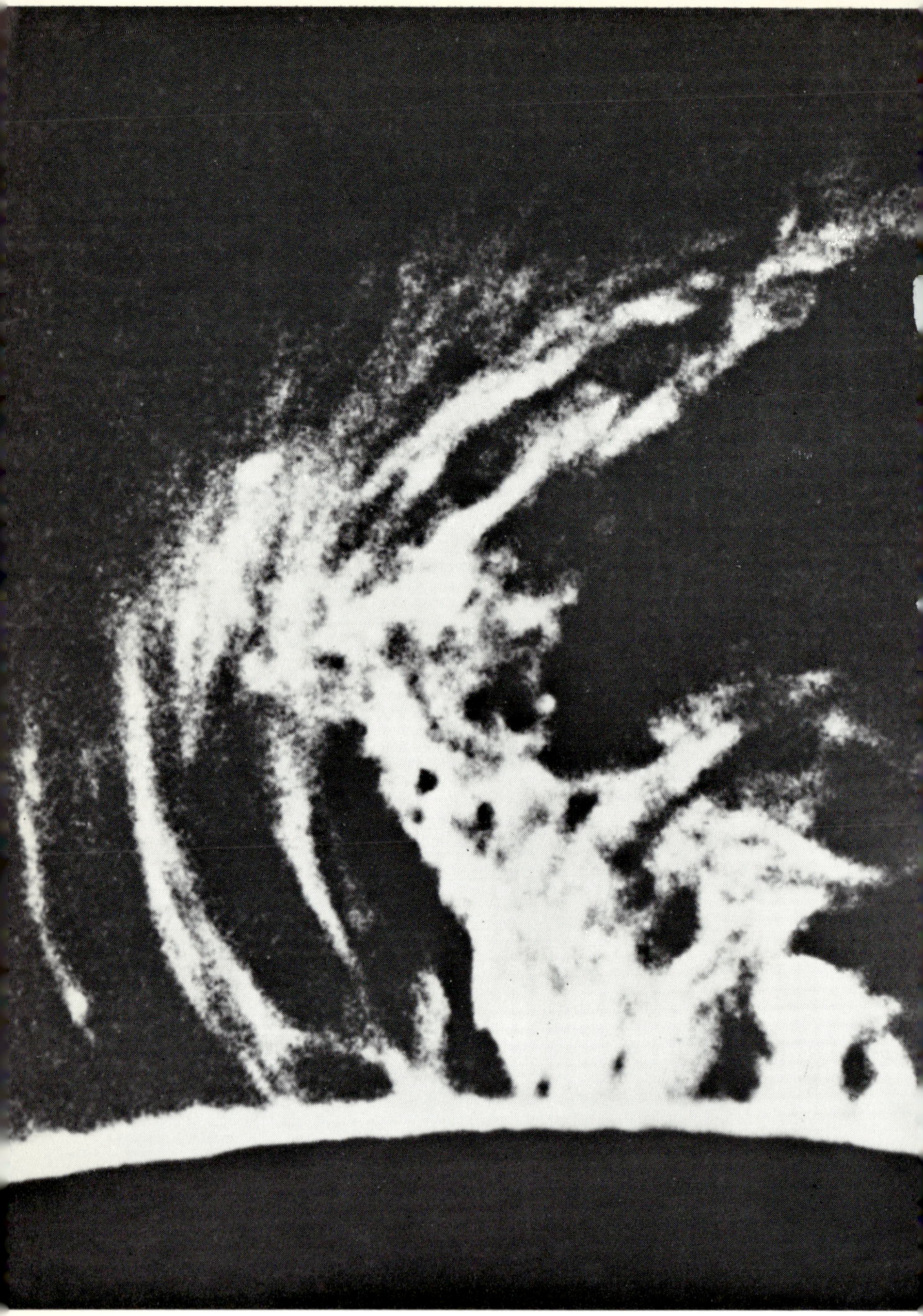

Plate 5. One of the most spectacular plasma displays is a solar prominence. The electrical and magnetic nature of plasma, which results from its composition of free positively charged particles and free electrons mixed together, is shown by the way the prominence follows magnetic field lines at the Sun's rim.

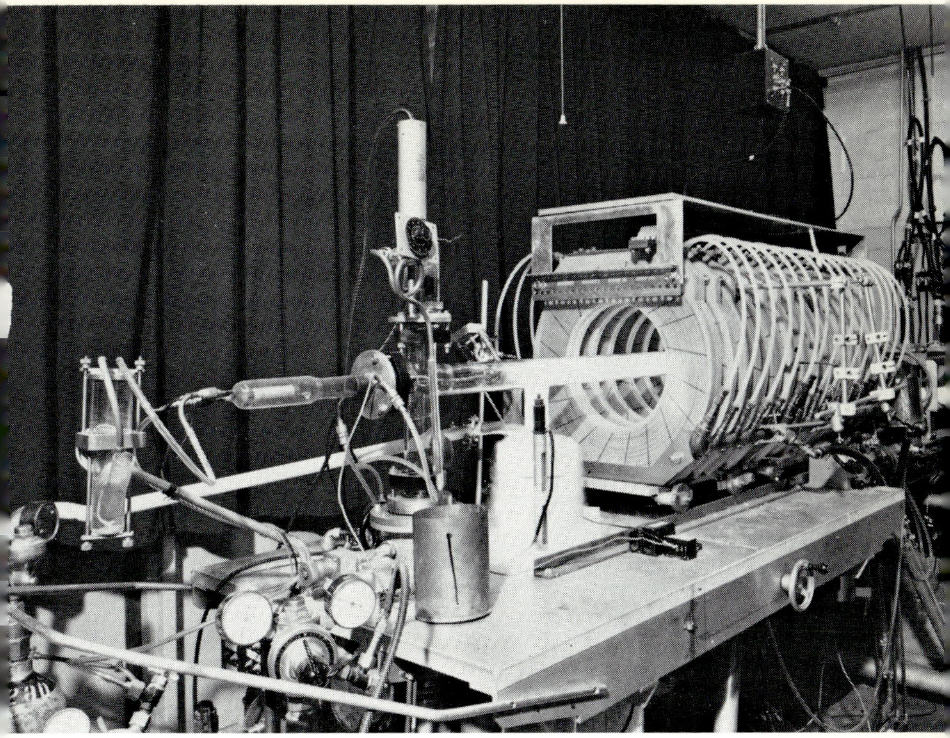

Plate 6. In the laboratory, plasma can be created by heating a gas to a very high temperature — above 5 000°C. In the apparatus shown a plasma column has been generated to study wave motions in the column.

Plate 7. The profiles of radio wave intensity (cf. 'brightness') of the radio source 'Cassiopeia A' have been produced by a computer from observations of the source made with the 'one mile' radio telescope at Cambridge. The 'mountains' represent points in the source emitting very intense radio signals.

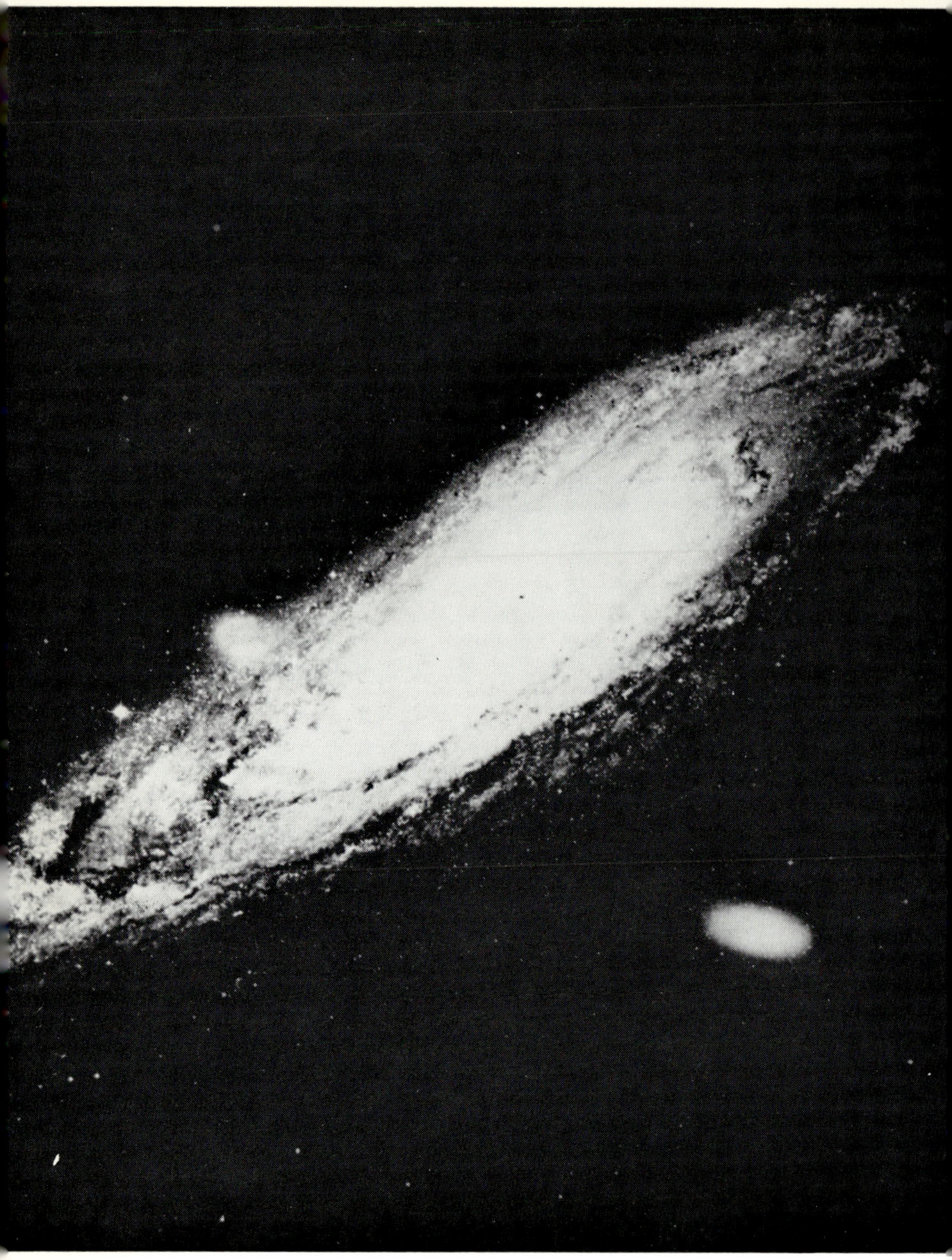

Plate 8. A galaxy of some 100 000 million stars in the constellation Andromeda. It is similar to our galaxy in structure and is 9×10^{18} miles away. There are at least two other galaxies in the picture, which illustrates how a 'Big Bang' could distribute galaxies throughout the Universe.

Plate 9. The 5-kilometre radio telescope, which is now being built at Cambridge, comprises four fixed and four movable 60 ft diameter aerials mounted along a former straight section of railway line. In use the telescope has the resolving power (i.e. ability) to generate detail of one 5-kilometre diameter aerial. It should show the fine detail in the most distant radio sources.

Plate 10. The tube, through which pulses of 28 000 million electron volt protons are fed into one of the Intersecting Storage Rings at Geneva, is shown going off to the right from the proton synchrotron ring at CERN.

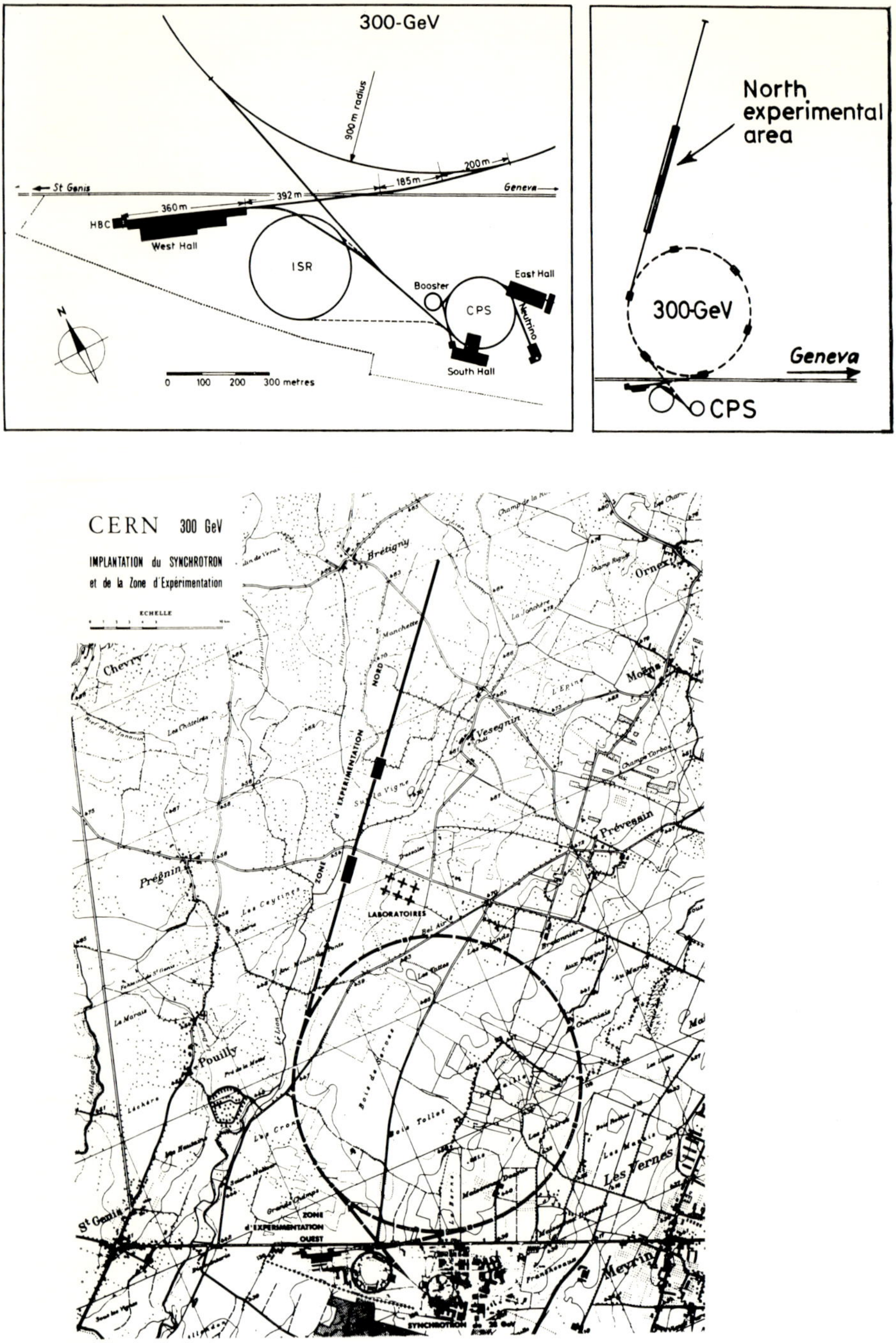

Plate 11. The layout of Europe's projected 300 GeV proton synchrotron. The maps and diagrams give some idea of the size of the projected new machine compared with the existing 28 GeV machine and Intersecting Storage Ring system.

known force, the amount of matter present could be determined by the acceleration acquired by the piece of matter. If a body of mass is suspended a short distance d above the Earth mass M, then the force acting upon m to pull it down (and the force acting upon M to pull it up) would be given by $F = \frac{G\,mM}{(D+d)} = mp$, where p would be the acceleration that m would acquire if it fell. D of course is the radius of the Earth. From this equation we see that $p = \frac{GM}{(D+d)^2}$ which is constant and independent of the mass of the falling body so long as d is very small compared with D, that is one or two thousand centimetres compared with 630 000 000 centimetres. This shows that all bodies accelerate at the same rate near Earth irrespective of their masses, which is what Galileo found out from the experiments he is said to have carried out at the leaning tower of Pisa.

If a force is applied to a mass and if the force is known and the acceleration can be measured, then the mass can be calculated. The process is more usually put to use the other way round, that is in the sense that if the mass is known, the acceleration can be calculated if the force is known or the force calculated if the acceleration is known. The former set of circumstances is the one usually applying in space flight where the mass to be accelerated is known, as is the force. What the astronauts want to do is to calculate the velocity their crafts will attain when the rocket thrust is applied to accelerate or decelerate it. In other circumstances it is the force that has to be found, for example, in an aircraft crash, when the investigators want to find out what stresses the structure was subjected to.

There are other properties of matter which might be used to measure mass, but in the main they are not so used, as the means

FIG 19
Gamma radiation from a radioactive isotope suffers a reduction in intensity in passing through a metal sheet. The reduction is proportional to the thickness and density of the sheet—the greater the thickness or density the greater the reduction. The thickness of sheet metal—say steel—can be monitored by comparing the reduction in intensity of gamma rays caused by the production line steel with that caused by a standard sample of sheet.

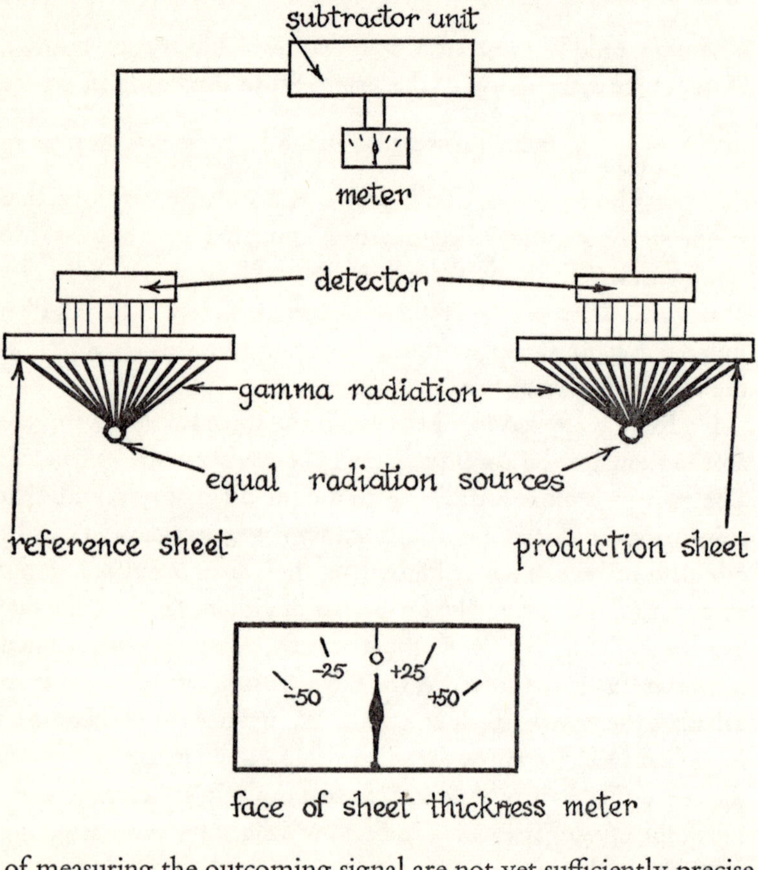

of measuring the outcoming signal are not yet sufficiently precise compared with commoner methods. For example, when electromagnetic radiation of certain kinds is directed at matter, it passes through the matter and becomes weakened as a result. The degree of weakening, or attenuation as it is called, suffered by a particular kind of radiation is proportional to the amount of matter placed in its path. So lead or platinum or uranium of a given thickness would produce a much greater weakening than would a corresponding thickness of aluminium or magnesium or paper (see Fig. 19). This property is not used to determine the mass of a leg of lamb in the butcher's in the High Street, but it is used very widely to control the thickness of uniform materials that are produced in sheet form, e.g. paper, plastics, steel, aluminium. For very thin, light materials, the radiation used is beta rays (electrons) produced by particular radio-isotopes, for thicker and heavier materials, like sheet steel up to 10 cms thick, the effective radiation is gamma rays again produced by specific radio isotopes (Plate 4).

The same sort of process is undergone with sound waves, they are more strongly attenuated by dense, massive materials than they are by light ones. This property is not however used for measuring mass but rather the massiveness of lead and concrete is used to stop sound.

To summarize the methods that have been described for measuring mass, the one that finds the commonest use is that depending on the property of matter to attract other matter. This not only gives us the standard method for determining any mass in commerce, trade, the household or the market place, but it also enables anyone to determine the mass of distant planets, stars and galaxies if they are within range. Other methods depending on inertia and the property of matter to attenuate radiation also find their use in special circumstances, but none are so widespread as gravity.

But let us take a closer look at the indestructibility of matter and mass.

4 The conservation of mass

At the end of the eighteenth century, when chemistry began to be an exact and well-understood science, scientists, using concepts of the atom and all that they meant, were able to carry out experiments to check these concepts, to test their validity. They found that if they weighed out reactants with the utmost precision of which they were capable, caused the reaction to take place under such conditions that they could control the whole process without loss, and then recovered the products and weighed them, the sum of the weights of the reactants was equal to the sum of the weights of the products, Substituting 'mass' for 'weight', this result always meant that within the limits of experimental error, mass was conserved in all the experiments carried out, even though in some of these experiments considerable amounts of heat had been absorbed or given out. This result we would say today, with considerable hindsight, would be a reasonable one.

Later in the nineteenth century, when physicists and engineers had got round to examining processes in which heat and other forms of energy were generated or transformed one into another or used to produce work, the conclusion was ultimately reached that energy was also conserved. That is if one measured all the energy put into a process and all the energy outputs from that process, the two sums would be equal. Energy incidentally is the property of a body or machine that makes it capable of doing work. A stone that is held by a string a hundred feet above the earth has the capability of doing work if the string is cut and the stone allowed to fall to earth. This can perhaps be better understood by considering water. We know that water trapped in a reservoir high in the hills is capable of driving a large turbo-generator to create considerable amounts of electricity that can heat our houses, drive our trains and the machines in our factories. The water in the reservoir has this 'potential' energy, the energy of position, which has to be converted to another kind of energy, 'kinetic' energy, the energy of motion, before it can be used to generate electrical energy. But in such a process, the conversion of 'potential' to 'kinetic' to 'electrical' energy, a balance can be struck and one can account for all the energy throughout the process. Starting with a given amount of potential energy, which can be calculated from a knowledge of the mass of the substance that is to move, then the amount of kinetic energy that will appear, the amount of electrical energy that will be generated and the amounts of energy that will be lost in the processes in friction, as heat and as noise, all these can be accounted for and a true balance struck.

Thus it came to be recognized that energy, as well as mass, was also conserved. In fact, following Einstein and his General Theory of Relativity, there is only one thing that is conserved and this is energy. Mass only appeared to be conserved because the instruments available for measuring it were insufficiently delicate to detect the discrepancy in mass that would occur when energy (heat or light) was evolved or absorbed in a chemical reaction. Speaking with hindsight this was not surprising. One outcome of Einstein's deliberations was his by now famous equation relating mass and energy

$$E = mc^2 \text{ or } m = \frac{E}{c^2} \text{ where } c = \text{velocity of light} = 3 \times 10^8 \text{ metres per sec.}$$

which enables one to calculate how much energy a given amount of mass can be converted into. In an ordinary chemical reaction, the amount of energy released or absorbed is relatively small and the discrepancy in mass too small to be measured. For example if the complete combustion of one gramme of coal yields $3{\cdot}344 \times 10^4$ joules, then one tonne of coal (10^6 grammes) if completely burnt would yield $3{\cdot}344 \times 10^{10}$ joules. This quantity of heat energy is equivalent to 0·37 milligrammes, a quantity that could readily be measured as such, but which is unlikely to be detected by a balance designed to weigh one tonne to an accuracy of one part in 1 000 million (0·37 milligramme is less than one thousand millionth part of one tonne).

It only became possible to detect discrepancies in mass balances, that is in equations in which the sum of the masses on one side should equal the sum of masses on the other, when it became possible to carry out nuclear as opposed to chemical reactions. For example, one kind of uranium will combine with a neutron and then split into a number of pieces:

$$U^{235}_{92} + n^1_0 \rightarrow U^{236}_{92} \rightarrow Mo^{99}_{42} + Sn^{135}_{50} + 2n^1_0 + \text{Energy}$$

the sum of the masses on the left hand side is 236·1265 atomic mass units (amu) while the sum of the masses on the right side is only 235·9165 amu. The discrepancy is 0·210 amu which is equivalent to an awful lot of energy, about 200 Mev (see footnote at end of chapter). This is why the atomic bomb is so powerful and why one ought to be able to extract from one tonne of uranium as much heat as from 3 million tonnes of coal. Another classic case is the energy released in the reaction in which hydrogen is converted into helium, the process that takes place in the stars. It is responsible for their ability to generate an enormous amount of heat and radiate it out into space for millions of years. The basic process is the building of helium nuclei out of hydrogen nuclei and neutrons (to avoid the complication of dealing with electrons) according to the following equation:

$$2H_1^1 + 2n_0^1 = He_2^4 + \text{energy}$$

The masses on the left add up to 4·033262 atomic mass units while the mass of the helium atom that is created is 4·003874 atomic mass units. The discrepancy on 4 mass units is nearly 0·03 compared with 0·21 on nearly 240 atomic mass units which means that much more energy can be got from a given mass of hydrogen than can be got from the same mass of uranium and this is why the hydrogen bomb is so much more powerful than the atom bomb. This process accounts for the amount of light, heat, etc., that the Sun, a not very big star, radiates out into space every second, something it has been doing for thousands of millions of years, a matter of some 4 million tonnes.

But what we have been leading up to is an explanation of the strange statement, made in Chapter 3 (page 46) that the behaviour of matter would lead one to know how much mass was present 'at least with reference to the place and circumstances in which the measurement was made'. As mass can change into energy, shown in the examples quoted above, so energy can be changed into mass. In high energy physics these processes are happening all the time. Particles such as hydrogen and other nuclei (atoms stripped of their outer electrons) are accelerated in atom smashing machines such as the big proton synchrotrons at CERN in Geneva. In such a machine a bunch of protons (hydrogen nuclei) would be accelerated to an energy of say half a million electron volts in a linear device, one that accelerates particles in a straight line. The energy acquired would be equivalent to the energy the particle would have attained as energy of motion had it been injected at a metal plate at zero potential (0 volts), and been attracted to another plate at 500 000 volts. At this energy, the bunch of particles would be admitted to the circular race track of the proton synchrotron itself in which they would be constrained by magnets to move in a circular orbit, to circle the track 50 000 or 60 000 times, to receive a voltage kick of some 50 000 volts every time round and to end up at an energy of 30 000 000 000 electron volts moving at a speed more than 99% of the speed of light. Now when a particle reaches a speed of 9/10ths the speed of light, its mass has more than doubled and the faster it goes the faster its mass increases. As the mass tends to increase faster than the speed, there comes a point at which virtually all the energy added turns into mass and there is practically no increase in speed. This is why the biggest accelerators get to be known as ponderators.

You may well ask at this point, 'how does one know that fast particles have more mass than slow ones?' The answer comes from the kind of device that is used to make the tracks of charged particles visible, the development of the cloud chamber, the liquid hydrogen bubble chamber. In this device, if the liquid is at a temperature just above its boiling point and a charged particle passes through the liquid, then the ions created by the particle colliding with molecules in the liquid act as centres for bubbles of vapour to form in the liquid, so that the path of the passing particle is marked by a trail of bubbles (Fig. 20). These can be photographed.

FIG 20
A charged particle, passing through a superheated liquid (one that doesn't boil even though above its boiling point) will leave behind it a trail of invisible ionized particles each of which can act as a focus on which vapour bubbles can start to form. The minute initial bubbles constitute a sort of vapour trail behind the ionizing particle and its whole track can be recorded by photography using a short duration flash. Frequently liquid hydrogen is used to act as the target and the liquid; the tracks are recorded on photographs taken from three mutually perpendicular directions.

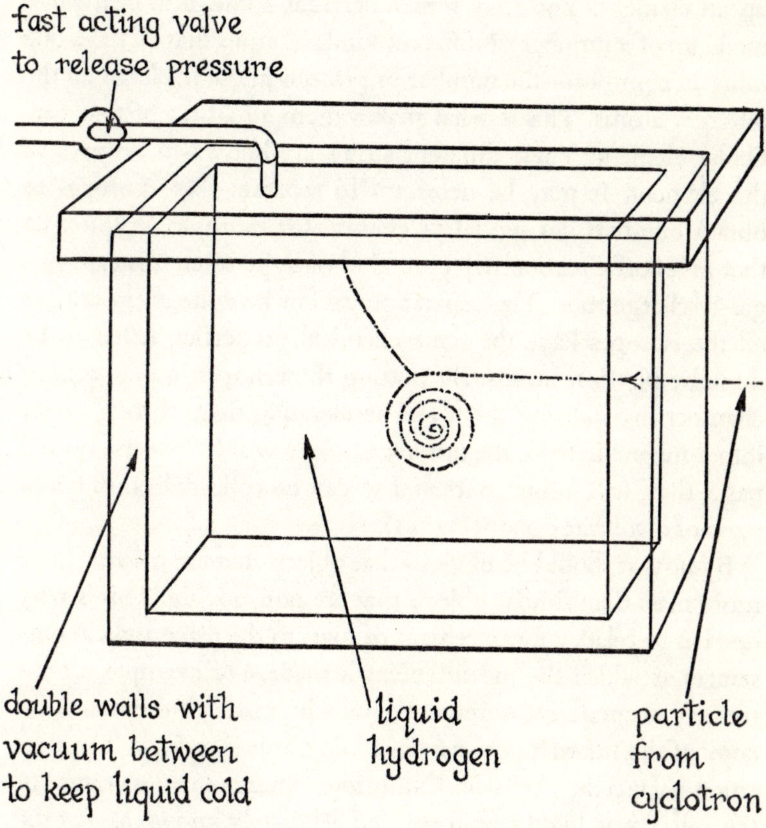

If a powerful magnetic field is created in a direction perpendicular to the path of the particle, the track will be curved more for a slow light particle than for a fast heavy one. By measuring these curved tracks and knowing something about the particle and the dimensions of the bubble chamber, the mass of the particle making the track can be determined.

This technique of shooting charged particles into a strong magnetic field is used to separate the different isotopes that make up an element. You may remember that a chemical element is made up of a number of different kinds of atom that all have one thing in common – the number of protons in the nuclei of all the different atoms. This is what makes them all atoms of that particular element. These different atoms are known as isotopes of the element. It may be necessary to separate these isotopes to obtain one that has special properties, for example, krypton 86 that gives off a particularly pure kind of light when it is used in a gas-discharge tube. This separation cannot be done chemically as all the isotopes have the same chemical properties, it has to be done by physical means. By putting the isotopes into a vacuum chamber, by ionizing them and accelerating them through a slit into a magnetic field, the heavier isotopes will follow less curved paths than will lighter ones and so can be collected on different parts of a collector plate (Fig. 21).

By now it should be obvious that objects that are moving have more mass than similar objects that are not moving. This is why one has to relate a measurement of mass to the place and circumstances at which the measurement is made. For example, an extremely accurate measurement might be made of the mass of a copy of the International standard kilogramme, using a high precision balance in a cellar in Cambridge where the temperature of the cellar was highly uniform and accurately known as was the

FIG 21
In a uniform magnetic field, a mass spectrometer, the paths followed by charged particles are curved. If two or more isotopes of different mass numbers are equally charged and injected at the same speed in the field of a magnet, the radii of the paths of the heavier particles will be larger than those of the lighter ones. In this way chemically identical isotopes can be separated.

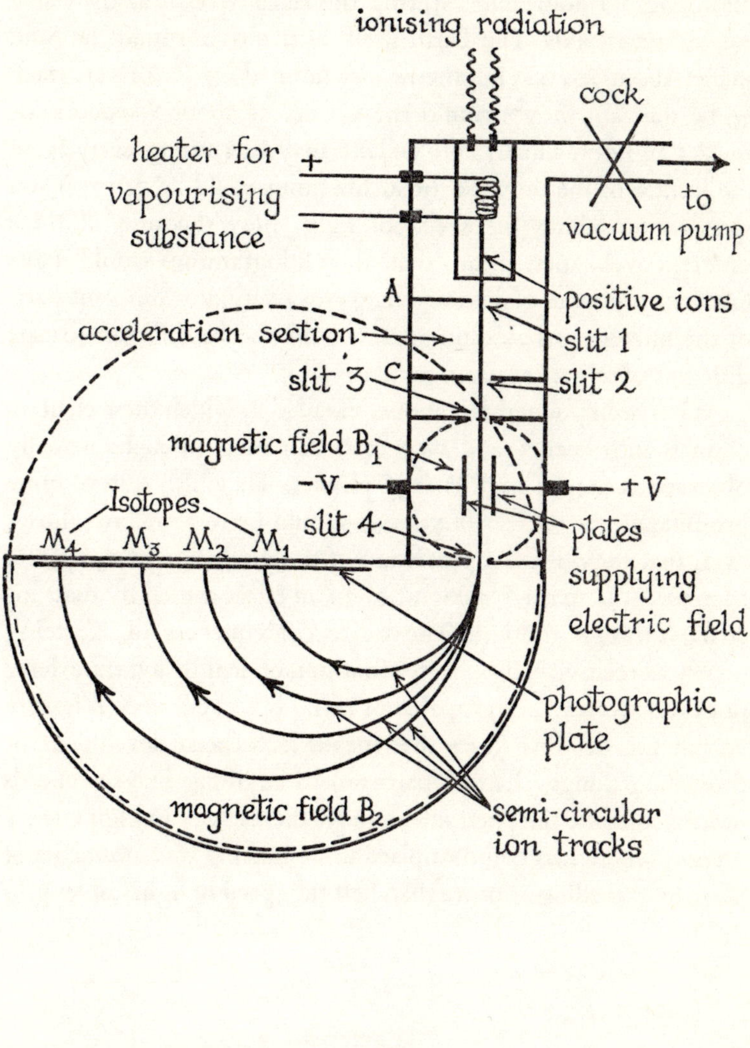

barometric pressure. So, on the face of it, it would seem perfectly reasonable to say that the mass of the copy kilogramme was accurately known and equal to that of the standard with which it was compared in Cambridge. Further one might say that one knew how much mass was there – after all the cellar was not moving – or was it? The Earth's surface was moving at 1 500 kilometres an hour, and carrying the cellar with it as the Earth rotated on its axis. The Earth itself, as it moves round the Sun, makes about 100 000 kilometres per hour along its orbital track while the Sun moves round the Galaxy at about 800 000 kilometres an hour. Finally, the Galaxy may be rushing away from the 'centre of the universe' (if such a point could be defined) at a speed approaching the speed of light. Now because all these different velocities mean that our kilogramme would have different amounts of kinetic energy in relation to different parts of the universe, its absolute mass would also be different for the different observers at those points.

As has been pointed out above, the field in which these changes of mass into energy and vice versa can most often be actually observed is that of high energy physics. They also feature quite prominently in nuclear physics. It should be remembered, however, that the observed precession of the perihelion of the planet Mercury was an early recognition of an effect caused by mass increase through motion. The other consequences of Einstein's theory of relativity, the effects of motion on length and time, have not until recently been capable of being observed, since it is only within the last few years that means have been developed for propelling things (charged particles) to anything like the speeds needed to make the phenomena show themselves. In high energy physics where it is commonplace to be making measurements of particles travelling at more than half the speed of light, all lengths and times derived from bubble chamber photographs are automatically corrected for relativistic effects. It is interesting to note the effect that relativity has on the half lives of the mesons, the nuclear particles having rest masses of between one tenth and one half of a proton rest mass, which are produced when fast protons hit stationary ones. In bubble chambers many such stationary particles are produced and it is comparatively easy to observe how long they live before they decay into electrons and energy quanta. From the various lives observed, the half life of a particular kind of meson can be calculated and used to predict how many of this kind of meson should survive a high-speed journey through a long, evacuated flight tube and arrive at the other end. On account of the change in the time scale due to the speed at which the mesons travel, a high percentage of the speed of light, many more survive the journey than the number predicted by the rest half life, and the increase is what would be expected according to Einstein.

In the 1880s the American physicists, Michelson and Morley, attempted to find out if the velocity of light was lower in the direction the Earth was moving than it was in the direction at right angles to this direction. The velocity should be lower if the Earth (and everything else) were moving through a sea of 'ether', a medium that the physicists felt they needed to exist to explain how energy could be transmitted over the vast distances of space. This medium was peculiar in that it was thought to be able to pass unimpeded through matter. It is not, we can see now, such an odd property as it first sounds when you consider that matter is largely empty space. But in these circumstances, if light is a wave motion in the 'ether' and if the 'ether' flows though the Earth, you would expect the velocity of light to be lower 'against the stream' as opposed to across it. Michelson and Morley set up an elaborate

FIG 22
The Michelson-Morley experiment. The luminiferous ether, the medium needed according to nineteenth-century scientists for the propagation of light, etc. either drifted along with the Earth or drifted through its atoms. The experiment shown was designed to detect motion of the ether relative to the Earth by comparing waves sent along the direction of the ether current and across it. The experiment showed no significant motion.

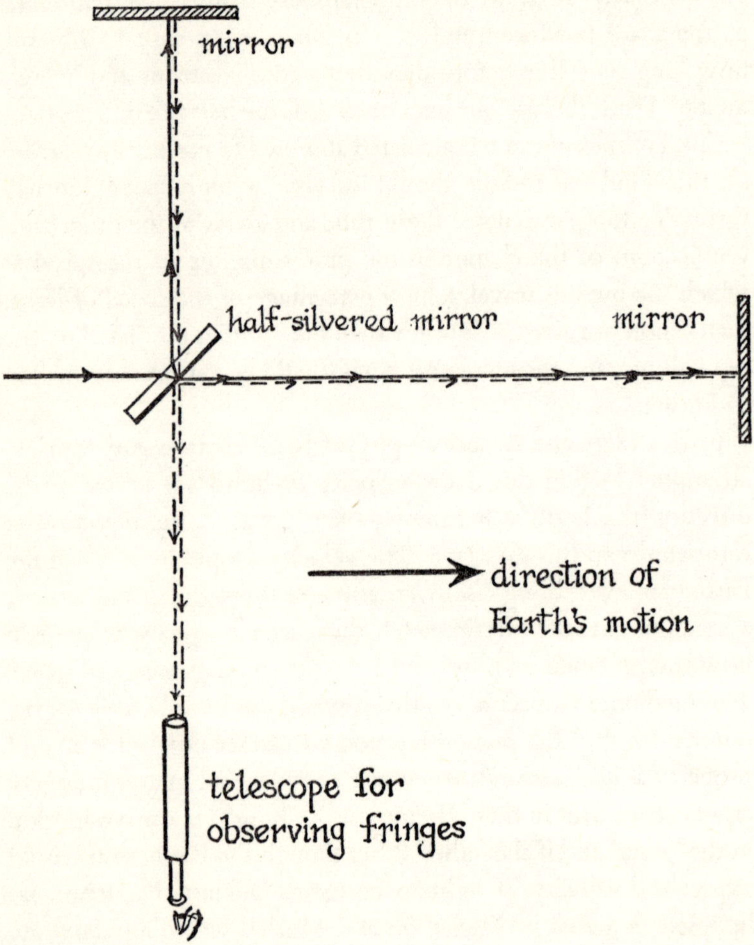

experiment to try to demonstrate the movement of the Earth through the 'ether'. It consisted of an optical system mounted on a massive stone slab which was floated in a tank of mercury to insulate the apparatus from vibrations and to make it easy to turn through various angles. In the optical system, a beam of light was split by passing through a half silvered mirror, one half going straight on to travel a fixed distance in the direction that the Earth was travelling through space, the other half being sent off at right angles to travel the same distance at right angles to the direction of the Earth's motion. Both beams, at the end of their journeys, were reflected back along the paths they had come to be reflected or transmitted through the half-silvered mirror to form an interference pattern in a telescope (Fig. 22). Now had there been any effect on the velocity of light by any motion of an 'ether' relative to the Earth, Michelson would have expected to find a fairly large displacement of the fringe system set up in their telescope. This they were unable to do and it had to be concluded that there was no movement of the 'ether' relative to the Earth. This result was confirmed by many subsequent experimenters.

The Irish scientist, Fitzgerald, suggested that the reason for this result was that in the above experiments, the path in the direction of the Earth's motion actually contracted by an amount that would exactly compensate for any increase in the time of flight of a light ray in this direction. He further proposed that it would be impossible to detect this contraction because any measuring rod introduced to measure the contraction would suffer in the same way. This, broadly, is the same conclusion that was reached by Einstein, who showed that velocity not only affected mass and length but also time, changes in all three becoming significant as velocities approached the velocity of light, the ultimate velocity that could not be exceeded. As a body approached in velocity

FIG 23

The Earth's gravity pit. The Energy needed to move unit mass from a point distant from the Earth's centre to infinity is given by

$$\text{Energy} = E = \text{the integral from r to infinity of } \frac{R^2}{r^2}\,dr.$$

$$= \frac{gR^2}{r} \text{ where } R = \text{radius of earth. As R and g are constant,}$$

E is proportional to 1/r. This is shown in the curve which plots E against r. E = gR = energy to move body from Earth's surface to infinity.

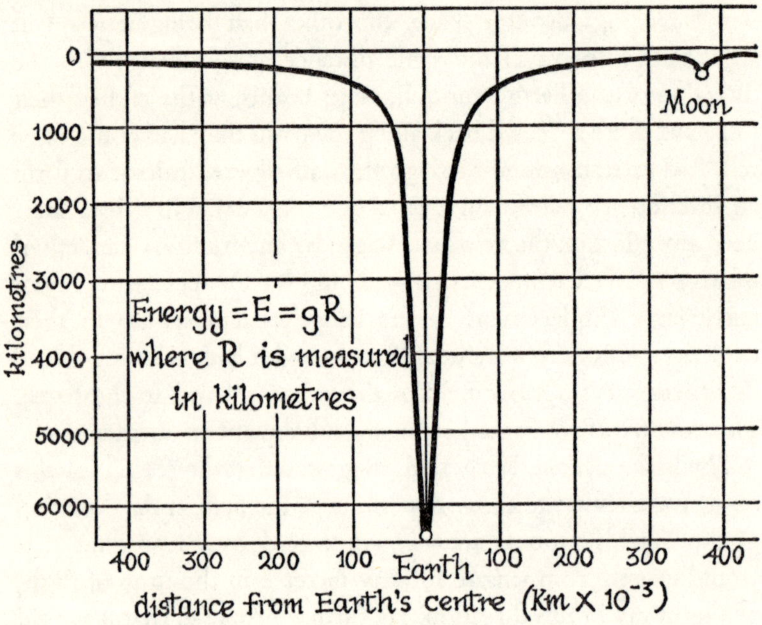

closer to this ultimate, so its mass would tend to increase towards infinity and its length in the direction decrease towards zero. There would also be major disturbances in the times of flight. These matters can be observed in experiments with high-speed particles, as has been said before, but for man to be able to take part in such an experience himself, the art of rocketry has got to advance quite a lot from its present state. At the beginning of the

1970s rockets are capable of lifting three men in their life support capsule and their exploration vehicle with the third stage of the launch vehicle off the ground, into orbit about the Earth and finally on to an escape path which takes them to the moon. The three-stage rocket system, the first stage of which develops about 3½ million kilogrammes of thrust, can lift the payload out of the Earth's gravity pit (Fig 23) and send it on its way at some 40 000 kilometres per hour. If the system missed its target, the moon, for any reason, the payload would probably go into orbit round the Sun perhaps reaching high speeds at some parts of its journey. But the rockets available now cannot store sufficient energy to make really high speeds attainable, and by this we mean speeds that are an appreciable fraction of the speed of light which is more than 1 000 000 000 kilometres per hour. Rockets powered by atomic engines are being developed which should enable much higher speeds to be attained but there is still a long way to go before man himself can experience (if indeed someone undergoing the process can be said to experience it) the process of mass addition by kinetic energy or length contraction of any appreciable magnitude.

Footnotes

[1] If the masses in the Uranium-neutron equation were expressed in grammes, the discrepancy would be 0·210 grammes

$= 0{\cdot}210 \times 0{\cdot}6 \times 10^{24}$ a m u

$= 0{\cdot}210 \times 0{\cdot}6 \times 10^{24} \times 4{\cdot}15 \times 10^{-17}$ kilowatt hours

$= 5$ million kilowatt hours

[2] 1 a m u $= 1{\cdot}6603 \times 10^{-24}$ grammes $= 4{\cdot}15 \times 10^{-17}$ kilowatt hours. Therefore 1 gramme = 25 million kilowatt hours.

[3] The conversion of hydrogen into helium in a star such as our Sun is believed to take place by the carbon-nitrogen reaction outlined below:

$$C^{12}_{6} + H^{1}_{1} = N^{13}_{7} + \text{Energy}$$
$$N^{13}_{7} \rightarrow C^{13}_{6} + e^{0}_{1}$$
$$C^{13}_{6} + H^{1}_{1} = N^{14}_{7} + \text{Energy}$$
$$N^{14}_{7} + H^{1}_{1} = O^{15}_{8} + \text{Energy}$$
$$O^{15}_{8} \rightarrow N^{15}_{7} + e^{0}_{1}$$
$$N^{15}_{7} + H^{1}_{1} = O^{16}_{8} + \text{Energy}$$
$$O^{16}_{8} \rightarrow C^{12}_{6} + He^{4}_{2}.$$

Nett process

$$4\,H^{1}_{1} \rightarrow He^{4}_{2} + 2\,e^{0}_{1} + \text{Energy}$$

5 More about matter

In conclusion, something more must be said about matter which is really what this book is all about. It seems as if matter (and hence all the material things we know, including ourselves) is made out of a lot of different pure substances which in turn are made out of a substance which may be a sort of condensed energy. This is how it appears, as in certain circumstances it can turn into pure radiant energy, whereas in other circumstances pure radiant energy can be transformed into the substance that is the basis of matter. The ninety or so basic substances, the elements, from which every known substance and all those at present unknown can be constructed, all contain atoms which are built on a simple solar system plan. In this, each atom contains a nucleus, in which the greater part of the mass of the atom is concentrated, surrounded at a great distance by a number of orbiting electrons which, by their various gyrations, form the surface of the atom and its character as we know it. These electrons contribute little to the mass of the atom but a great deal to its physical and chemical properties, to its hardness or softness, its colour, taste and so on.

Let us go in greater depth into the last point made, by assuming that we could make ourselves grow smaller and smaller and, simultaneously, move closer and closer to a bright, shiny, salmon pink cube of pure copper standing on a flat piece of glass on the ground. If we could assume that our eyes would get more and more capable of seeing small things as we got smaller and smaller, then we would first be able to see the crystals in the block of copper, then the atoms in the crystals (Fig. 24). This could not happen in fact because our eyes (even if very tiny) can only work with visible light radiation and this sets a limit to the size of things that can be defined, however . . . If we had made the cube of copper with a side 10 centimetres long, then we would know that we were dealing with a mass of 8 950 grammes (why?) and at the start, when we were still quite big, the side of the block would appear as a shiny, smooth, pink cliff with clean sharp edges. Along each of these edges, some 44 million atoms would be deployed but long before we were small enough to be able to see atoms, we would have lost sight of the edges of the cube and we would not be in a position to count heads.

When we had reduced our size to the level at which the atoms in the copper cube were bigger than we were, we would then be confronted with a wall of atoms, each one vibrating in its appointed place in the crystal lattice. What we would see of each atom would be its outer skin, which we might imagine as a four-layer transparent shell made of ceaselessly orbiting electrons, twenty-nine in all. This would be the outermost boundary of the atom, the part like the skin of an orange, responsible for 'the general impression' that we have of the atom in a physical sense. If we could see through this boundary, which contributes only a tiny amount to the mass of the atom, we might be able to pick out the nucleus sitting at the centre of the whole arrangement. But although the nucleus contains virtually the whole mass of the atom, physical measurements suggest that its diameter is only

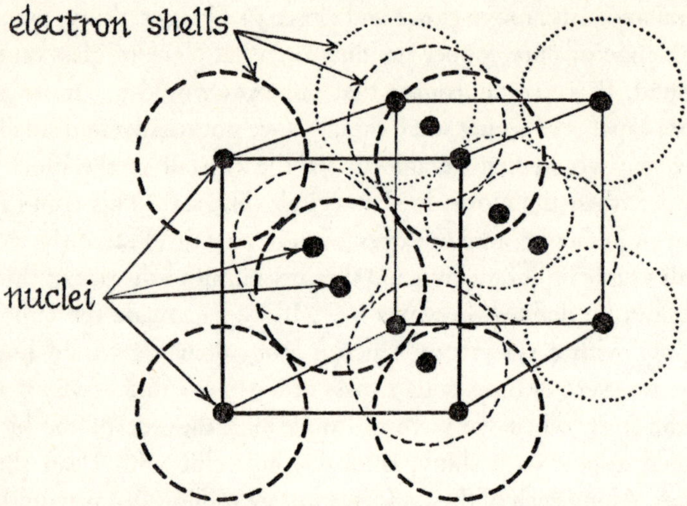

FIG 24
The copper crystal unit is a face centred cube, a system which has one atom at each corner of a cube and an atom at the centre point of each face. The atoms in the front face, which actually 'touch' each other, are heavily outlined for clarity.

about one ten-thousandth part of the diameter of the whole atom. So, if the nucleus appeared to us to be about half a centimetre in diameter (the size of a pea), we, standing at the boundary about 75 metres away, would hardly be able to see it (fig. 4).

This tiny speck, made as it would be out of the fundamental 'matter stuff', would be pretty remarkable for a number of reasons. In the first place it would be very much denser than anything we might meet in the every-day world. Density is the property that tells us how much matter there is in a standard volume of stuff. For example one cubic centimetre of lead, which is known to be 'heavy', contains 11·7 grammes compared with 1·0 grammes for one cubic centimetre of water. For comparison, the masses of one cubic centimetre of other dense elements are:

uranium 18·7 grammes, tungsten 19·3 grammes, platinum 21·45 grammes, and osmium 22·5 grammes (which is nearly twice the density of lead). To get an idea of how dense the 'fundamental matter stuff' must be, let us consider one cubic centimetre of copper. This mass would contain 86 000 million, million, million atoms and hence 86 000 million, million, million nuclei. The latter would be responsible for very nearly the entire mass of 8·95 grammes. But the volume actually occupied by all the nuclei together would be far less than that occupied by all the atoms together which we know to be one cubic centimetre. The volume occupied by the nuclei should be about one ten-million-millionth of a cubic centimetre so that the density of the 'fundamental matter stuff' should be about 100 million million grammes per cubic centimetre.* This means that a matchbox full of the

* If every kind of nucleus is in fact made out of fundamental matter stuff, we would expect the nuclear radius to be proportional to $A^{\frac{1}{3}}$, A being the number of nucleons in the nucleus, the mass number, and hence a measure of the volume of the nucleus. Empirically the nuclear radius, r_a, is given by $r_a = 1{\cdot}5 \times 10^{-13}.A^{\frac{1}{3}}$ cms, hence the radii of the nuclei of H, Be, Al, Cu, I, and Bi are approximately as follows:

H^1	$= 1{\cdot}5 \times 10^{-13}$ cms	Cu^{64}	$= 6{\cdot}0 \times 10^{-13}$ cms
Be^8	$= 3{\cdot}0 \times$ do.	I^{125}	$= 7{\cdot}5 \times$ do.
Al^{27}	$= 4{\cdot}5 \times$ do.	Bi^{206}	$= 9{\cdot}0 \times$ do.

If the density of any element is s grammes per cm^3, then the number of atoms in s grs.

$$= N.s/A \text{ where } N = \text{the Avogadro } N^0 = 6 \times 10^{23}$$

Whence total vol. of all these atoms

$$= \tfrac{4}{3}\pi \cdot r_a^3 \times N.s./A \text{ cm}^3$$

Inserting the value of r_a, and of N, the total volume, V,

$$= \tfrac{4}{3}\pi \times 3{\cdot}375 \times 10^{-39} \times 6 \times 10^{23} \times s \text{ cm}^3$$

Whence the density of fundamental matter stuff

$$= S = \frac{s}{v}$$

$$= \frac{s}{\tfrac{4}{3}\pi \times 3{\cdot}375 \times 6 \times 10^{-16} \times s} \text{ gr/cm}^3$$

$$= 10^{14} \text{ gr/cm}^3$$

'fundamental matter stuff', alone would weigh about 3 500 million tonnes, a truly fantastic weight.

But atomic nuclei, containing 'fundamental matter stuff', are remarkable in many other ways. For example they are all able to hold together very stably, most of them have been around for more than 4 000 million years, and keep bound up within their walls a lot of electrically charged particles that normally repel one another very strongly. Another remarkable thing is the complexity that lies within even the simplest nucleus, that of hydrogen.

Atoms can be considered to be made of three fundamental particles – protons, neutrons and electrons. The first two of these, known as nucleons, are similar in size and mass. The proton, which is the nucleus of the hydrogen atom, is positively charged and very stable, the neutron is uncharged and unstable, when not confined within an atomic nucleus. When free, the neutron decays radioactively into a proton and an electron and has a half life of only 13 minutes (this means that if one had 1 000 neutrons at one instant then 13 minutes later only 500 would remain unchanged). The third particle, the electron, is much lighter than the other two with a mass little more than one two-thousandth that of the proton. Its charge is opposite in sign but equal in size to that of the proton, and the electron is usually to be found, with many others, in orbit round a nucleus. The nucleons are held together by a very strong force field which manifests itself as mesons of various kinds when a nucleus or a nucleon is broken up in a collision. The force field may also appear as additional mass attached to a nucleon.

Atoms of any one particular element all have nuclei having one thing in common, they all contain the same number of protons. The atoms of the element may differ in that they may have different numbers of neutrons. Thus a number of varieties of an

FIG 25
Isotopes are atoms of an element which all have the same number of protons in the nucleus but which may have different numbers of neutrons. This is shown diagrammatically.

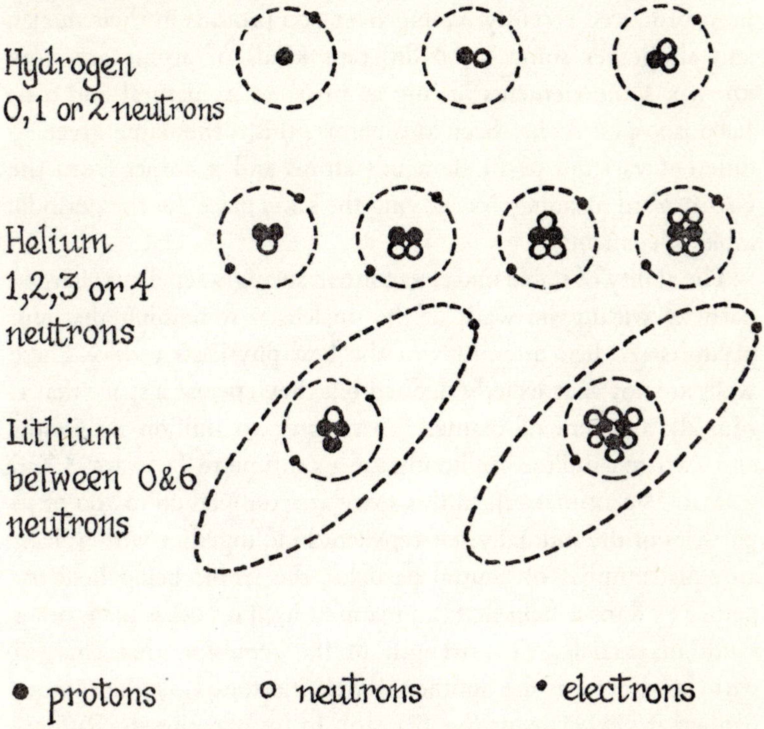

element may exist, these varieties are called isotopes. Hydrogen, the lightest and simplest element, has atoms in which there is always one proton. Its nuclei can have 0, 1 or 2 neutrons. This gives a range of three kinds (isotopes) of hydrogen which are physically quite different but chemically very similar. The next lightest element, helium, has two protons in its nuclei with one, two, three or four neutrons and the one after that, lithium, three protons with one to six neutrons (see Fig. 25). There are 92 naturally occurring elements (the heaviest – uranium – having 92 protons in its nucleus) and each one differs from the one before by having one more proton in its nucleus. New elements have been produced recently having over 100 protons in their nuclei, and altogether some 1 300 different kinds of atoms are now known, some elements having as many as 25 natural and artificial isotopes. As has been said before, this is the name given to different varieties of an element's atoms and it comes from the Greek word meaning, 'occupying the same place' (in the periodic table of elements).

The ability of stable nuclei to contain strongly repellent charged particles within the walls of the nucleus is something that still occupies the best attentions of the best physicists today. These walls are not very exactly defined but they enclose a space that is roughly a sphere of diameter between two million millionths and two ten-million-millionths of a centimetre (2×10^{-12} and 2×10^{-13} centimetres). In this space are confined up to 100 or so particles of the mutually self-repellent kind together with at least an equal number of neutral particles, the whole being held together by a force field that can manifest itself on occasion as other kinds of particle. The strength of the repulsion that charged particles exert on one another is well illustrated by the famous Rutherford experiment (see Fig. 26). In his experiment, Rutherford caused alpha particles (helium nuclei shot out from radium at thousands of kilometres per second) to impinge on a thin sheet of gold foil. He observed that a few out of the many bullets fired, on approaching the flimsy sheet of gold, were turned directly back on their tracks. These were those alpha particles that had been directed by chance straight at the centre of a gold nucleus. This experiment not only served to define the dimensions of the nucleus it also indicated the speed that 'nuclear bullets' need to

FIG 26
The Rutherford experiment to demonstrate the emptiness of atoms. He fired alpha particles from Radium C at a thin sheet of gold leaf and found that although most particles passed straight through, some were deflected and a few turned back on their tracks. From his observations he was able to deduce not only the principles of the 'Solar System' atom but also its main dimensions.

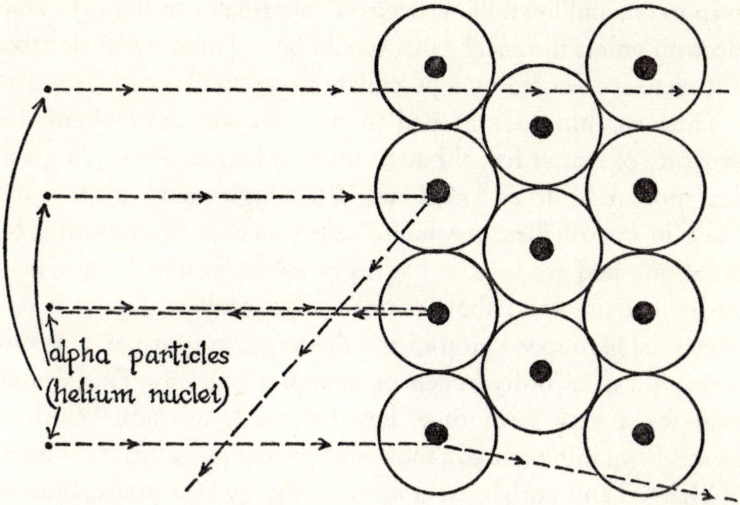

have if they are going to be successful in breaking through the electrostatic barrier surrounding the nucleus.

Experiments such as Rutherford's indicated a clear need for special machines for accelerating charged particles to high velocities for probing the secrets of the atom. Beginning with such machines as the Cockcroft and Walton cascade generator, which in 1930 was used to make the first artificial transmutation of elements, today scientists have developed machines which accelerate protons to energies of more than 25 000 million electron volts and to speeds very close to the velocity of light. The targets shot at are often hydrogen nuclei, in the form of liquid hydrogen, and sometimes heavier and more complicated nuclei. Three machines already exist which can endow protons with energies of between 28 000 and 70 000 million electron volts and plans are afoot to build even bigger machines to produce protons with energies of 200 000 to 400 000 million electron volts and perhaps even to one million million electron volts (Plates 10 and 11). (One electron volt is the energy that would be acquired by an electron if it were accelerated by a potential of one volt.)

These machines are needed to find out still more about the structure of matter for, although much is known already, a great deal more needs to be found out. When high energy protons are made in one of these machines, called incidentally proton synchrotrons, and are fired at hydrogen atoms in liquid form in a device known as a bubble chamber, the result of collisions between the high speed protons and the target protons, may be the formation of a 'di-hydrogen or helium 2' nucleus. This would exist for a very short time, less than 10^{-24} seconds, before it broke down into perhaps a shower of pi-mesons, which are lumps of charged and uncharged condensed energy with rest masses of about 270 to 280 times that of an electron. Alternatively, the breakdown products might be energetic protons and neutrons, covered with lots of condensed energy and with rest masses more than 20% greater than those of normal particles. These are known as lambda, sigma and xi particles and they only appear in special circumstances.

Professor Weisskopf of the European Centre for Nuclear Research (CERN) has summarized the process of finding out about the structure of the atom. He pointed out that to establish the planetary nature of the atom, energies of only a few electron volts were needed to make the atom reveal that it had a nucleus surrounded by clouds of electrons orbiting in specific shells. Much higher energies, of the order of one million electron volts, were needed to smash the nucleus of an atom, a large proportion of the energy being needed to make the probe particle go fast enough to crash through the repulsive field barrier that surrounds the nucleus. Once this barrier is surmounted, the strong repulsion somehow seems to change into a very strong cohesive force holding the charged and uncharged particles together in the nucleus. One of the things discovered about nuclei, particularly the more complex kinds made up of considerable numbers of protons and neutrons, was the fact that in such nuclei the nucleons (protons, neutrons) were arranged in shells in a manner somewhat similar to that of the atom's electrons in their courses, but perhaps more like the skins of an onion. Neutrons and protons were found to be changing one into the other and back again, alpha particles (helium nuclei) were known to assemble themselves from neutrons and protons and exist for a brief period of time, of the order of 10^{-20} seconds or less, before dispersing and going back into the merry-go-round. This particular phenomenon has a bearing on the relative infrequency with which alpha particles are emitted from some radioactive heavy elements which would be expected to be unstable from a knowledge of their compositions, elements like uranium and thorium which have half-lives of millions of years.

The next stage in the process described by Professor Weisskopf is that of probing the structures of the nucleons themselves, the protons and neutrons, and this calls for machines capable of accelerating the probing bullets to energies of at least several thousand million electron volts (several GeV). The current range of experiments, carried out with the 28 to 70 GeV machines referred to earlier (p. 82) which incidentally include cyclotron-type machines for producing very high energy electrons for use in nucleon probing experiments, have yielded a large amount of information on the structure of nucleons. They have also shown that there is a large and growing family of 'anti particles', particles that are in a sense, mirror images of the particles with which we are now familiar. The first of these particles to be discovered was the positron, the positive electron, which is identical with the ordinary or negative electron except that it has a positive charge, equal in size and sign with that of the proton. Positrons are comparatively rare in this world because, with so many electrons about, a positron inevitably meets an electron after only a short period of free existence. When this happens both are annihilated and replaced by two photons of radiant energy, two gamma rays. Conversely when two such photons meet they can be converted into a positron/electron pair. If such an event occurs within a magnetic field, the positron will follow one path curved in one direction, the electron will follow a different path curved in the opposite direction.

Very recently other anti particles have been discovered, mostly in experiments in large accelerators, and these now include the anti proton, the anti neutron, the anti alpha particle (helium nucleus) and most of the anti mesons and baryons. The latter are heavy particles, as their names suggest, with rest masses greater than those of the nucleons. All this leads physicists to suppose that whole galaxies, maybe whole universes, may exist which are composed of anti matter. These would be indistinguishable from galaxies made of what we think of as ordinary matter until the two kinds happened to come into contact. They would then annihilate each other and disappear in a colossal gust of radiant energy.

It is to be hoped that the new generation of high energy accelerators may throw more light on the question—what is the fundamental matter stuff?' They may even provide the answer. It is already suspected that its density must be even higher than the figure calculated from nuclear dimensions (see page 77) since there is strong evidence to suggest that the nucleus contains discrete bodies made of something very dense *and* a lot of empty space. The new accelerators may also help to solve other problems connected with the nature of matter. For example, is it feasible to consider that hydrogen atoms may be created spontaneously in intergalactic space at a definite rate that can be calculated? This concept, which is not all that outrageous in view of the fact that positron/electron pairs are known to be created out of the fusion of two photons of energy, played an important part in the 'Steady State' theory of the origin of the Universe. This theory suggested that the Universe always appeared to the observer as it appears to us today. If the Universe is expanding and galaxies gather speed the further they get from the 'centre', they must ultimately attain and exceed the velocity of light, when they would no longer be able to send back information to us, the observers. They disappear over the 'speed of light horizon'. So with galaxies disappearing, there was a need for new ones to appear if the observable universe was to appear to remain the same. Stars, and presumably galaxies, were known to condense out of the intergalactic thin soup of hydrogen. But this hadn't enough body to supply new galaxies at the rate required to keep the system apparently constant, particularly as space itself was expanding and so reducing the concentration of hydrogen. So, to fulfil the requirements of theory, it became necessary to postulate that hydrogen be created spontaneously at a rate sufficient to sustain the hydrogen concentration needed. It turned out that this rate was pretty low, so low as to be very difficult to detect. Experiments carried out to detect the creation have so far proved inconclusive. All the same, despite its shortcomings, the Steady State theory was very satisfying philosophically as it implied that the Universe would have no beginning and no end.

The alternative hypothesis, known as the 'Big Bang' theory, assumed that the expanding universe resulted from a sort of super-super nova explosion that took place around 10 million years ago when all the matter and energy in the Universe was concentrated into one compact and relatively small super galaxy or even star. This would be 'The Beginning'. As a result of the explosion, great concentrations of matter, chunks almost, would have been sent flying in all directions at different energies. These masses, mostly hydrogen, would then start developing and evolving into the galaxies as we know them today. Since the 'Beginning', they have been accelerating away from the origin and have now attained considerable speeds and reached considerable distances from it. Some may long since have exceeded the speed of light and vanished over 'the speed of light horizon'. But all the galaxies we are in touch with today can be studied in the state of development they were in when the communicating radiation—light, radio waves, ultra-violet rays, X-rays, etc.—set out across space. The radiation from the galaxies at the edge of space has been over 3 000 million years reaching the Earth, even though travelling at 300 000 km. per second, so we are 'looking' at them as they were 3 000 million years ago. Undoubtedly some of them have by now disappeared over the visible horizon, but even so the astronomers can study stars and galaxies in a wide variety of states of development (Plate 8).

Thus the 'Big Bang' theory was supposed to begin with a 'Beginning'. The end of the Universe would occur when every galaxy had been isolated from its fellows by the 'speed of light' barrier, when all the intergalactic/hydrogen had been used up so that no new stars or galaxies could condense out, and when all the stars had burnt themselves out. On balance, the evidence that has been accumulated tends to support the Big Bang theory. This evidence is connected with the distribution of rapidly receding galaxies at the edge of the universe.

Today, astronomers and radio astronomers like Sir Martin Ryle at Cambridge and Sir Bernard Lovell at Jodrell Bank incline to a 'Big Bang' theory that embraces not a once-and-for-always event but a series of Bangs that take place every 20 000 or 30 000 million years. We are now in an expansion phase, following a 'Big Bang' which took place some 10 000 million years ago. Some force, perhaps gravity, will have to slow down and stop the outward flying debris and then cause it all to start accelerating in again until it is all once more assembled in the super galaxy and ready to start the expansion process again. Long before the next bang takes place, all the universe we knew would have disappeared. A whole new universe would be created in the aeons following the explosion and life, which would have been wiped out in the creative process, would have to begin again in different forms and in different places.

Important results, having a material bearing on the nature of matter, have been announced by the radio-astronomers in the last few years. These have come from the discoveries of quasars and pulsars, two remarkable phenomena in the Heavens (Plate 7). Quasars, or quasi stellar radio sources, look like nearby stars of moderate brightness but are in fact extremely distant bodies of extraordinarily high brightness. They are so bright, so extremely energetic that an understanding of them calls for new developments in physical theory. No doubt the high energy physicists will have a hand in this. Pulsars are sources of radio waves that emit pulses with extreme regularity, at intervals ranging from

1/30th to 4 seconds and their regularity has been compared to that of the best of clocks. They are thought to be neutron stars and represent the final stages in the evolution of a star, i.e. collapse into a body of very high density (see page 78), only a few kilometres across. In such a regime, where very high gravitational forces are operating, only neutrons can exist, all other atomic structures must apparently break down. The collapse process may be associated with a super nova type event which leaves behind a body spinning at such a high speed that it has to oscillate between a ball and a doughnut configuration. Such an arrangement could account for the nature of and the regularity of the radio pulses. Pulsars may be the origin of cosmic rays. If they were to collapse even further and end up with a diameter of only a few metres, raising the density to an even higher level than the 10 million million grammes per cubic centimetre that they are now thought to have, a new and interesting situation would arise. The gravitational force associated with such a mass concentration would be so high as to prevent the escape of any radiation whatsoever. In these circumstances, the star could disappear completely from the ken of outside observers who depend on receiving information, radiation, to know that something is there. These stars would disappear just as effectively as those galaxies that are thought to have passed over the 'speed of light' barrier. The difference would be that some future space traveller could suffer considerable embarrassment were he to come into the vicinity of such a dark star, he is unlikely to run across an 'over-the-barrier' galaxy.

To reiterate, the new accelerators and astronomical instruments are likely in the future to provide so much more information about problems associated with matter that we shall inevitably gain a better understanding of its nature (Plate 9). It is possible that one very significant property, gravitation, the weakest of weak forces, may come to be understood. Weak though it is, it is obviously of great importance to planet-bound creatures. Should it ever be explained and understood, the possibility of controlling it would become feasible. Men of the future would then have space travel on tap and have no further need for those tremendously thirsty engines – rockets.

Index